TRENDS AND PERSPECTIVES IN PARASITOLOGY: 2

6

TRENDS AND PERSPECTIVES IN PARASITOLOGY

2

EDITED BY

D. W. T. CROMPTON AND B. A. NEWTON

CAMBRIDGE UNIVERSITY PRESS

CAMBRIDGE

LONDON NEW YORK NEW ROCHELLE

MELBOURNE SYDNEY

Published by the Press Syndicate of the University of Cambridge
The Pitt Building, Trumpington Street, Cambridge CB2 1RP
32 East 57th Street, New York, NY 10022, USA
296 Beaconsfield Parade, Middle Park, Melbourne 3206, Australia

First published 1982

Printed in Great Britain at the
University Press, Cambridge

Library of Congress catalogue card number: 81-642963

British Library Cataloguing in Publication Data

Trends and perspectives in parasitology. – 2.

1. Parasitology – Periodicals
574.5′249′05 QL757

ISBN 0 521 24830 2 hard covers
ISBN 0 521 28989 0 paperback
ISSN 0260-6763

CONTENTS

PREFACE

The first volume of *Trends and Perspectives in Parasitology* (TRAPS) offered six articles that dealt with the traditional research territory of the parasitologist. The articles focussed attention on developing and growing points in parasitology, identified new and old problems in need of research and presented readers with concise summaries of recent discoveries. In planning the second volume, we invited a marine biologist and a palaeontologist to write about their researches in the hope that readers of TRAPS might learn about results from other disciplines which could have some bearing on their own investigations. Whether information about the clownfish–anemone symbiosis or the fossil record will stimulate those working on the evasion of the immune response or the evolution of parasitism remains to be seen. We hope, however, that all our readers will enjoy and be intrigued by this collection of essays which appeared in *Parasitology* in 1981. Finally, it is with much pleasure that we thank the seven contributors whose efforts have produced this volume.

Cambridge, December 1981

D. W. T. Crompton
B. A. Newton

CONTRIBUTORS

A. J. CLARKE graduated in chemistry from Glasgow University and was awarded the Ph.D. degree in Organic Chemistry from Imperial College, London, in 1955. He is now Principal Scientific Officer in the Department of Biochemistry at Rothamsted Experimental Station where he is particularly interested in the chemicals involved in nematode hatching.

S. CONWAY MORRIS graduated in geology at the University of Bristol before obtaining his Ph.D. degree from Cambridge in 1976. His research in palaeontology encompasses the importance of soft-bodied faunas in the geological record, the biological and evolutionary events around the Precambrian–Cambrian boundary and co-evolution in the fossil record with particular reference to parasites. He is presently a lecturer in the Department of Earth Sciences at the Open University.

SHEELAGH LLOYD qualified in veterinary medicine from Trinity College Dublin in 1969 and was later awarded the Ph.D. degree in immunoparasitology at the University of Pennsylvania. She is now working as a Senior Research Associate at the Department of Clinical Veterinary Medicine, University of Cambridge, where she is studying immune responses to metacestodes and immunological unresponsiveness to gastrointestinal helminths in pregnant and lactating hosts.

H. R. LUBBOCK was killed in a road accident in Brazil in 1981. His diverse research activities included studies on the taxonomy of fishes from coral reefs and the biology of sea anemones. The latter interest, which he described in this volume of TRAPS, prompted him to begin an investigation into mechanisms of cellular recognition. With the tragic death of Roger Lubbock, biology has lost a most able experimentalist and all who knew and worked with him send their sympathy to his family and friends.

GROVER C. MILLER studied zoology at the University of Kentucky and Louisiana State University before moving to North Carolina State University at Raleigh where he is Professor of Zoology. He is chiefly interested in the ecology and life history of helminths of wild vertebrates of the southeastern United States. He has served as President of the Southeastern Society of Parasitologists and the Association of Southeastern Biologists.

R. N. PERRY received his Ph.D. degree in Zoology from the University of Newcastle-upon-Tyne in 1975. After carrying out postdoctoral research at Newcastle, he moved to the University of Keele for 3 years before joining the staff of Rothamsted Experimental Station where he is a Senior Scientific Officer. In addition to his research on nematodes, he contributes to courses at Imperial College, London, and the University of Reading.

M. L. SOOD graduated from Lucknow University where he also studied for the Ph.D. degree. He is now Associate Professor in Zoology at Punjab Agricultural University where, in addition to his interest in nematode morphology and ecology, he is investigating the metabolism of *Haemonchus contortus*.

Parasitology (1981), **82**, 159–173

The clownfish/anemone symbiosis: a problem of cellular recognition

ROGER LUBBOCK

Department of Zoology, University of Cambridge, Downing Street, Cambridge CB2 3EJ

(*Accepted* 30 *July* 1980)

INTRODUCTION

Sea anemones, jellyfish and other cnidarians are characterized by the presence of stinging cells known as nematocytes. Each of these cells contains a minute, often venomous, harpoon which is discharged explosively upon contact with prey. Nematocytes function semi-autonomously and have a sophisticated ability for recognizing foreign animals. Surprisingly, there are a few unusual animals which can live in close contact with cnidarians without ill effects.

This review examines the mechanisms that have been proposed to explain how these symbiotic animals can survive in such a potentially lethal environment. Special attention is paid to the symbiosis between clownfishes (*Amphiprion*) and sea anemones which has been the subject of recent experimental work (Lubbock, 1979*a*, 1980).

ASSOCIATIONS WITH CNIDARIANS

An indication of the phylogenetic diversity of animal species that have been found in association with cnidarians may be obtained from the studies of Allen (1972), Bruce (1976), Hayes & Grayson (1968), Mariscal (1974) and Rees (1967). Unfortunately, most of the work concerning animals that are symbiotic (symbiosis is used here as originally defined by De Bary, simply to indicate the living together of unrelated organisms; see Mariscal (1971)) with cnidarians has been primarily taxonomic and, in the majority of cases, it is unclear whether these animals do or do not elicit the discharge of nematocysts (the nematocyst is the explosive organelle contained within each nematocyte).

In some cases initial work has indicated that the symbiotic animal does not induce discharge, although no attempts have been made to examine the mechanisms responsible for the apparent protection. Thus, Gorton (1960), Fricke (1967) and Nayar & Mahadevan (1967) reported that shrimps (*Periclimenes*) and crabs (*Petrolisthes*) lived among the tentacles of giant tropical sea anemones without appearing to be stung but they did not provide experimental evidence for their conclusions. Briggs (1976) demonstrated experimentally that the copepod *Paranthessius anemoniae* was not stung by its host anemone *Anemonia sulcata*, although the reasons for its protection were unclear. In addition, he found that the copepod possessed a certain resistance to anemone toxins in solution. Ross (1974) reviewed some of the extensive work carried out on the hermit crab/anemone symbiosis and it is interesting to note that, apart from a few observations, little is known about the protection of the hermit crab from the nematocyst discharge of its resident anemones. Similarly, certain protozoa associated with *Hydra* do not appear to cause discharge of nematocysts, although the mechanisms responsible have not been studied (see Hayes & Grayson, 1968). Rützler (1971) reported on an unusual genus of burrowing sponges (*Siphonodictyon*) whose embryos apparently settle on living coral tissue. He found that the embryos were provided with large amounts of mucus and suggested that this mucus protected the larva from the coral polyps during settlement.

Wright (1858) first commented on the presence of undischarged nematocysts in eolid nudibranchs. More recent work (Graham, 1938; Edmunds, 1966; Thompson & Bennett, 1970; Harris, 1971; Mariscal, 1974; Conklin & Mariscal, 1977; Day & Harris, 1978) has shown that nudibranchs are able to absorb certain kinds of nematocyst selectively from their cnidarian prey and subsequently use them for their own defence. A particularly graphic example of this phenomenon was given by Thompson & Bennett (1970) who correlated the fact that bathers in New South Wales were quite severely stung by pelagic nudibranchs with the presence in these creatures of *Physalia* defensive nematocysts. The mechanism by which nudibranchs are able to ingest and store selected nematocysts in their cnidosacs without eliciting discharge has not been studied in detail. The most likely explanations proposed to date are those of Graham (1938) and Harris (1971). Graham suggested that the copious secretion of mucus, which is a normal preliminary to feeding, might make the mollusc less vulnerable to nematocyst discharge and also, that the nematocysts might be enveloped in the nudibranch's mucus without exploding. Harris suggested that the nematocysts were immature when ingested, and only matured in the cnidosacs.

In the case of the planarian *Microstomum*, Martin (1908) and Kepner & Nuttycombe (1929) have shown that the animals are able to store (and probably use as a defense) undischarged nematocysts obtained from feeding on hydroids. *Microstomum* must possess some mechanism, as yet uninvestigated, which enables it to swallow and subsequently retain nematocysts without causing them to discharge.

Certain fishes are particularly conspicuous cnidarian associates and have attracted a comparatively large amount of attention as regards the means by which they prevent themselves being stung by their hosts. Abel (1960) reported that the Mediterranean goby *Gobius buchichii* was able to live among the tentacles of *Anemonia sulcata* without being stung, its protection being restricted to the single anemone species. Physical damage to the fish was shown to cause a temporary loss of protection. Abel concluded that the factor responsible for protection from the anemone's nematocysts resided in the fish's skin. These results are analogous to those of Dahl (1961), who showed quantitatively that the skin of young whiting elicited virtually no discharge of *Cyanea* nematocysts, in contrast to the skin of a goby. Mansueti (1963) and Thiel (1979) reviewed researches concerning fishes associated with jellyfishes and provided circumstantial data, for the most part in agreement with those of Dahl (1961), concerning the fishes' protection from nematocyst discharge. Schlichter (1970) stated that the Indo-West Pacific wrasse *Thalassoma amblycephalus* occasionally swam among the tentacles of *Stoichactis* sp. (? = *Radianthus ritteri*) without being stung but he did not perform any detailed experiments. Hanlon & Kaufman (1976) reported on Caribbean fishes found in the close vicinity of sea anemones. Several of these did not seem to have protection from nematocyst discharge and avoided direct contact with the tentacles, but 5 species belonging to the clinid genera *Malacoctenus* and *Labrisomus* appeared to be able to move unharmed among the tentacles.

Among the many Indo-West Pacific members of the family Pomacentridae (damselfishes) are certain striking species which are known to live in association with sea anemones: these fishes belong to the genera *Dascyllus*, *Premnas* and *Amphiprion*. Of the 9 known species of *Dascullus*, only 2 (*D. trimaculatus* and *D. albisella*) have been found in association with sea anemones; the remainder (excluding the poorly known *D. strasburgi*) are usually found in live coral heads and have never been seen in sea anemones (Allen, 1975; Randall & Allen, 1977). All members of the genera *Amphiprion* and *Premnas* appear to be symbiotic with anemones (Allen, 1972, 1975). Under natural conditions these fishes are known to inhabit approximately 13 species of anemones belonging to 5 genera (Allen, 1972).

THE CLOWNFISH/ANEMONE SYMBIOSIS

Earlier work

The symbiosis between clownfishes (*Amphiprion*) and sea anemones is one of the few cnidarian associations that has been studied in any detail. It was first noticed over a hundred years ago by Collingwood (1868); other earlier workers were Sluiter (1888), Horst (1902), Verwey (1930) and Moser (1931). More recently, a wide variety of mechanisms has been proposed to explain the protection of these fishes from their host anemones.

One group of workers (Gohar, 1948; Hackinger, 1959; Koenig, 1960; Blösch, 1961, 1965; Graefe, 1963, 1964) has suggested that the anemone inhibits or alters its pattern of nematocyst discharge as a result of either behavioural, physical or chemical stimuli from its symbiotic fish. Gohar (1948) considered that certain host anemones could recognize individual fishes by their mode of movement. Hackinger (1959) and Koenig (1960) indicated that the fishes were protected by their behaviour, without providing any field or experimental evidence. Blösch (1961, 1965) stated that the fishes' protection was not absolute and that it arose from the anemone becoming habituated to their presence. Several subsequent workers have been unable to repeat Blösch's results; Mariscal (1970*a*, 1971) critically reviewed Blösch's work and suggested that a misinterpretation due to lack of controls seemed likely. Graefe (1963, 1964) postulated that the fishes' protection was the result of 'the state of adaptation of the anemone to a particular level of tactile stimulation and to a particular size of the stimulated area'. The results of subsequent experiments by Schlichter (1968) and Mariscal (1970*a*) were not in agreement with this statement. Neither Gohar (1948), Hackinger (1959), Koenig (1960), Blösch (1961, 1965) nor Graefe (1963, 1964) provided detailed and controlled experimental evidence for certain of their results, and it seems possible that some of their conclusions may have been based on rather slim evidence. Fukui (1973) provided limited quantitative evidence that the periodicity of spontaneous tentacle contractions decreased in the presence of *Amphiprion*; on the other hand Fricke (1974) was unable to find any medium to long-term effects of *Amphiprion* on their host anemones.

Fishelson (1965) followed Buhk (1939) and Caspers (1939) in suggesting that the host anemones 'are not adapted to the catching of living fish, and that this is the basic condition for the association of the fish with them'. This claim is in opposition to the results of several other workers (see Mariscal, 1971).

The observations and experiments of Davenport & Norris (1958), Abel (1960), Eibl-Eibesfeldt (1960), Mariscal (1965, 1966, 1967, 1970*a*, *b*, 1971, 1972), Schlichter (1967, 1968, 1972, 1975, 1976) and Fukui (1973) have suggested that the anemone is the passive partner (although it produces mucus that could subsequently be utilized by the fish) and that the fish is somehow able to protect itself by altering its external coating as a result of contact with the anemone. Several workers (e.g. Davenport & Norris, 1958; Eibl-Eibesfeldt, 1960; Schlichter, 1968) have indicated that the substances causing protection from nematocyst discharge reside in the fishes' skin and/or mucus, the flesh of *Amphiprion* being readily stung by host anemones.

Davenport & Norris (1958), Schlichter (1967, 1968, 1975) and Mariscal (1970*a*) reported that if clownfishes were separated from their anemones for a few days, the fishes were stung by their original partners on reunion; this was independent of the presence in the respective anemone of other fishes, whatever their size or species. The isolated fishes could only re-enter their anemones without being stung after a period of acclimatization. The above authors concluded that separation induced changes on the part of the fish rather than on the part of the anemone; in addition, Schlichter (1972, 1975) concluded that fishes did not have innate protection, since they could not enter anemones without a period of acclimatization. On the other hand Fishelson (1965) could find no evidence for acclimatization. Fukui (1973) indicated that clownfishes were very cautious during acclimatization and that, although they 'start back repeatedly' from the anemone, they were not in fact being stung; she based this conclusion on microscopical examination of the fishes' body scales. Fukui used the above experiment, and also another involving counting discharged nematocysts on the scales of acclimatized and 'unacclimatized' *Amphiprion*, as evidence that the fishes themselves were producing a protective substance. Unfortunately her definition of unacclimatized fishes was rather vague: she stated that 'caution was paid in ensuring their separation from the anemone prior to experimentation'. Since no indication was given of how long her unacclimatized fishes had been isolated from anemones, it was not possible to tell whether the apparent protection of the unacclimatized fishes was due to a factor in their own mucus, or resulted from retention of protective substances derived from a previous anemone host.

Until recently the most widely accepted explanation for the protection of clownfishes from the nematocyst discharge of their host anemones has been that of Schlichter (1972, 1975, 1976). Schlichter (1972, 1976) carried out a series of experiments involving radiochemicals and isoelectric focusing, which seemed to show that substances derived from the anemone were present on the body surface of symbiotic fishes. He stated that 'substances with inhibitory qualities (protecting substances) are produced by the anemones themselves . . . the surfaces of these fishes (*Amphiprion*) are not impregnated with protecting substances. Adapted anemone fishes, neighbouring anemones of the same species and other "adapted" objects are coated with the inhibitory substances and thus do not induce nematocyte discharge.' From this he concluded that the principal difference between symbiotic and non-symbiotic fishes was behavioural: while non-symbiotic fishes avoided anemones, symbiotic species were prepared to tolerate being stung in the short-term (i.e. during acclimatization) and repeatedly contacted the anemone until finally they became coated in anemone substances and could enter the anemone unharmed. It should be noted that while Schlichter's (1972, 1976) results seemed to show that the fishes were coated in anemone substances, they did not prove either that these substances were protective or that the fishes themselves lacked protective compounds. Furthermore, no clear experimental evidence was provided for the statement that anemones produce substances that are nematocyte inhibitors. Schlichter's conclusions appear to have been based primarily on a synthesis of his own results and those of certain earlier workers on acclimatization.

It is clear from the above survey that a wide variety of interesting hypotheses has been proposed to explain how clownfishes can live unharmed in sea anemones. A problem in assessing the results of earlier workers on this symbiosis arises from the fact that they did not all use the same species of clownfish or anemone. It is thus possible that some of the differences between previous studies may be attributable to interspecific variations in the protective mechanism. Moreover, the taxonomy of the genus *Amphiprion* was in disarray until the study of Allen (1972), so that species were frequently misidentified; the taxonomy of the host anemones still remains problematic.

Experiments on Amphiprion clarkii *and* Stichodactyla haddoni

It seemed evident that initially the most satisfactory method of investigating the protective mechanisms of cnidarian symbionts would be to examine one association in detail. The symbiotic association between *A. clarkii* and *S. haddoni* was chosen because the animals were readily available and could be maintained under aquarium conditions. In addition, this particular symbiosis had already been subjected to a certain amount of experimentation (see above).

A. clarkii is a fish that is known to inhabit several different anemone species under natural conditions (Mariscal, 1971; Allen, 1972). Initially, 3 readily available host anemones, *Gyrostoma hertwigi*, *Radianthus ritteri* and *S. haddoni*, were compared in order to determine which would be the most suitable as an experimental subject. The anemones were found to differ significantly in the nature of their stinging response. In *S. haddoni* a large number of tentacles adhered to test objects, nematocyst discharge occurred over a wide area, nematocyst density was high and there was a marked level of adhesion. In contrast, *G. hertwigi* and *R. ritteri* adhered to test subjects with at most a few tentacles; nematocyst discharge occurred only over a limited area, nematocyst density was comparatively low and virtually no adhesive force was produced. Since the stinging response of *S. haddoni* was strong and readily quantifiable, this species was favoured in subsequent experiments. The experiments carried out on *A. clarkii* and *S. haddoni* are described below in condensed form and for further details see Lubbock (1979*a*, 1980).

The first stage of the investigation involved obtaining background information on the anemone's stinging response. The nematocytes clearly played a central role in the fish/anemone symbiosis and yet surprisingly, almost no information was available on the nature of organic compounds provoking nematocyst discharge in the host anemones. Experiments were therefore designed to determine which of the major categories of complex organic substances present in the mucus and epidermis of fishes would be most likely to excite nematocytes. This was carried out by touching *S. haddoni* with glass rods coated with a wide variety of proteins, glycoproteins, polysaccharides and lipids and then recording the anemone's responses. The results demonstrated that *S. haddoni* was able to distinguish between different organic substances. Overall, proteins and glycoproteins tended to produce the strongest nematocyst discharge, while polysaccharides and lipids were usually less stimulatory. Thus, of the substances present on the surface of a fish, those containing a proteinaceous moiety appeared to be the most likely to elicit nematocyst discharge. No simple basis could be determined for the manner in which the anemone recognized the different compounds, suggesting that the recognition system associated with nematocytes was of a rather complex nature.

The first hypothesis to explain the clownfish's protection to be tested was that of Buhk (1939), Caspers (1939) and Fishelson (1965), who indicated that host anemones were incapable of capturing fishes. *Stichodactyla* was presented with freshly dead fishes that were attached to a spring balance so that the anemone's adhesion could be measured. *A. clarkii* was found to induce minimal adhesion (mean adhesion 1 gram force = approx. $9{\cdot}8 \times 10^{-3}$ N), while non-symbiotic species of similar size were strongly adhered to (mean adhesion to each of the 6 test fishes varying from 186·6 to 399·2 gram force). The extent to which *S. haddoni* adhered to all but *A. clarkii* was remarkable. Often tentacles became detached from the anemone rather than loose their hold on the fish: up to 30 tentacles were seen to detach on to test fish during one presentation. Close examination of 1 fish after removal from the anemone showed that most of the fin membranes were torn or had disappeared, while the majority of scales were absent. The view that the inability of host anemones to catch fish was the basic condition for the clownfish/anemone association thus appeared to be incorrect. Although certain host species such as *R. ritteri* might not be able to catch living fish, it seemed clear that others such as *S. haddoni* would be quite able to capture any fish that swam among the tentacles in the same manner as a clownfish.

These observations indicated that the clownfish might somehow be altering the anemone's behaviour or nematocyst discharge. The fishes' protection had been entirely attributed to such a mechanism by Blösch (1961) and, although this hypothesis had not been accepted by most subsequent workers, limited evidence was available that clownfishes could alter the anemone's behaviour (Mariscal, 1970*c*; Fukui, 1973). The effect of *A. clarkii* on *S. haddoni* was assessed firstly by quantifying changes in the anemone's external morphology that resulted from the fish's presence and secondly by measuring the stinging response to gelatine-coated cover-slips both in the presence and absence of clownfish. The results of these experiments indicated that *A. clarkii* had only a limited effect on *S. haddoni*: whereas *S. haddoni* tended to be rather static when alone, those in the company of clownfishes were relatively mobile, although retaining the same overall appearance. In spite of these changes the fishes' presence did not seem to cause any reductions in the anemone's adhesive force or density of nematocyst discharge: the adhesive force/tentacle remained at about 1 gram force, while nematocyst density stayed in the vicinity of 10^4 capsules/mm^2. It thus appeared that the clownfish's protection was the result of a local phenomenon occurring at the fish/anemone interface rather than of a general inhibitory effect mediated by the anemone's nervous system.

Since *A. clarkii* could move unharmed amongst the tentacles of *S. haddoni*, although the latter's powerful stinging response was not impaired, it seemed likely that clownfishes differed from non-symbiotic fish species either in the quality or quantity of their external mucus layer. No data were available on the quantity of mucus produced by *Amphiprion*. A number of measurements was therefore carried out in order to discover whether *A. clarkii*'s mucus layer differed in thickness from that of closely related fishes not found in anemones. The thickness of the mucus layer on individual fish scales was measured by coating the scales with fine carbon particles. Then, by means of Nomarski optics (Nomarski, 1955; Padawer, 1968), the depth of the mucus could be determined. The results showed that, on average, the mucus thickness of *A. clarkii* (approximately 9–14 μm) was about 3 times that of *Dascyllus aruanus*, *Chromis caerulea* or *Paraglyphidodon nigroris* (approximately 3–4 μm). The latter 3 species, which are also members of the family of Pomacentridae but do not inhabit sea anemones, did not show significant differences in mucus thickness. There were no significant differences in mucus thickness between *A. clarkii* inhabiting anemones and *A. clarkii* that had been isolated from anemones for 5 months, thus suggesting that in all cases the observed mucus layer had been largely if not wholly produced by the fish itself.

An examination of the staining properties of goblet cells in the epidermis of *A. clarkii* and non-symbiotic species showed that the mucus produced by clownfishes was different from that of related species not found in sea anemones. *A. clarkii* mucus was found to consist to a significant extent of neutral polysaccharide, while the mucus of *P. nigroris*, *D. aruanus* and *C. caerulea* was comparatively acidic; in the case of *D. aruanus*, and possibly *C. caerulea*, strongly acidic groups in the form of sialic acid appeared to be present. The mucus of *D. aruanus* and *C. caerulea* also seemed to be weakly sulphated. Further comparison of the mucus of *A. clarkii* and *D. aruanus* by ultracentrifugation in caesium chloride density gradients provided additional evidence that they were not identical. The principal mucus component, consisting of highly glycosylated glycoprotein and possibly polysaccharide, differed significantly between the 2 species. It was thus clear that the external mucus layer of *A. clarkii* was not only thicker than that of closely related species not found in sea anemones, but was also of different composition. Although it was not evident to what extent these differences could be related to the fishes' abilities to inhabit anemones, these results nevertheless suggested that investigation of *S. haddoni*'s responses to isolated fish mucus samples might prove fruitful.

S. haddoni was presented with glass rods freshly coated in fish mucus, the latter having been collected as carefully and gently as possible in order to minimize contamination with epidermal cells (cf. Lubbock, 1980). The initial experiment involved testing the anemone's responses to mucus from *A. clarkii* and the non-symbiotic species *C. caerulea*, *D. aruanus* and *P. nigroris*. Samples from *A. clarkii* were found to be unreactive, while those from the other 3 species were very stimulatory and induced a strong stinging response. These results matched those obtained using newly dead test fishes and were essentially analogous to observations made on living fishes.

The possibility that the unreactive nature of *A. clarkii* mucus in the above experiment might have been a function of anemone substances adsorbed on to the fish was investigated by presenting *S. haddoni* with fresh mucus samples from *A. clarkii* that had been isolated from anemones for 3–8 months. The clownfish mucus was found in general to retain its inability to incite *S. haddoni*'s nematocyst discharge. Significant quantities of anemone substances were unlikely to have been retained on the surface of *A. clarkii* for 3–8 months following isolation, and so the lack of reactivity of *A. clarkii* mucus was thought to be the result of substances produced by the fishes themselves.

In order to test the hypothesis that clownfish mucus contained inhibiting substances (Schlichter, 1972), *S. haddoni* was presented with rods coated in mucus from both *A. clarkii* and the non-symbiotic *D. aruanus*. Combined samples of *A. clarkii* and *D. aruanus* mucus elicited a strong stinging response in almost all cases. This suggested that, if an active inhibitor of nematocyst discharge was present in *A. clarkii* mucus, its effects were relatively weak or specific and insufficient to prevent stimulatory elements in *D. aruanus* mucus eliciting heavy discharge. Attempts were also made to alter *S. haddoni*'s responses by denaturing mucus from *A. clarkii* and *D. aruanus* in a variety of different manners. Rods coated with mucus were heated to 100 °C or were immersed in acetone, ether or in a chloroform–methanol mixture. The results showed that denaturation did not produce a marked change in the anemone's pattern of response. Whereas *D. aruanus* mucus remained essentially stimulatory, that of *A. clarkii* remained more or less unreactive. Excitatory substances appeared to be present in *D. aruanus* mucus but not in *A. clarkii* mucus. The possibility that *A. clarkii* mucus was essentially excitatory but achieved protection from nematocysts by means of specific substances that either chemically masked excitatory compounds or inhibited nematocyst discharge seemed relatively unlikely. Any such protective substance would have had to have been unusually resistant to denaturation to account for these results.

The possibility that fishes could gain protection from stinging by coating themselves in anemone mucus (Schlichter, 1972) was investigated by presenting *S. haddoni* with *D. aruanus* mucus coated in *S. haddoni* mucus. The strong stinging response caused by *D. aruanus* mucus on its own was not found to be affected even if the rod coated in fish mucus was immersed in anemone mucus prior to presentation. It should be noted that after immersion in *S. haddoni* mucus, and prior to immersion in the aquarium containing the test anemone, the rod could be clearly seen to be coated in a layer of anemone mucus. Furthermore, it was found that rubbing rods coated with *D. aruanus* mucus against the column of the test anemone prior to presentation to the latter's tentacles did not cause a reduction in response, even though this treatment might have been expected to coat the rod in anemone mucus and to reduce the overall amount of *D. aruanus* material present. Thus, no evidence could be found that coating the mucus of a non-symbiotic fish with a thin layer of anemone mucus resulted in the former losing its stimulatory capacities.

The above work indicated that the clownfish produced a thick layer of mucus which was inert to the host anemone's nematocytes. These results were in apparent contradiction to those of Schlichter (1972, 1975, 1976), who indicated that the clownfish's immunity from nematocyst discharge was due to a layer of protective substances derived from the anemone mucus. Although his experimental evidence was not conclusive, his assertion was worthy of further investigation. Radioactivity was incorporated into a specimen of *S. haddoni* using tritiated glucose such that its mucus became labelled. Experiments were carried out to test how much anemone mucus was transferred on to *A. clarkii* when in the anemone's company. In addition, *A. clarkii*'s affinity for anemone mucus was compared to that of related fish species that do not inhabit anemones in order to test for any special ability to take up anemone mucus. It was found that radioactivity on *A. clarkii* inhabiting *S. haddoni* tended to be higher than that on control fishes not allowed to contact the anemone, suggesting that animals in contact with *S. haddoni* could under certain circumstances acquire anemone mucus on their surface. Nevertheless, the amount of radioactivity that was transferred on to the fishes inhabiting the anemone was low; if one assumed that all radioactivity was derived from mucus and was evenly spread, then it represented a layer measuring about 0·01–0·28 μm in thickness. This was very thin when one considered that *A. clarkii*'s own mucus layer was 9–14 μm thick. Previous results had shown that non-symbiotic fish mucus could not be protected from stinging, even when coated with a visible layer of anemone mucus. It therefore seemed very unlikely that such a small amount of anemone mucus could be the prime factor responsible for *A. clarkii*'s protection. In addition, no differences could be found between *A. clarkii* and related non-symbiotic species in their affinity for labelled anemone mucus; this was perhaps a further indication that the anemone mucus on the clownfish was not of primary importance. No radioactivity was detectable on *A. clarkii* that were retested 24 h after their removal from the labelled anemone, although the fishes' protection persisted for months.

In summary, *S. haddoni* had a powerful stinging response which was sufficient to capture a small fish; this response did not seem to be markedly affected by the presence or absence of *A. clarkii*. *A. clarkii* achieved protection from stinging by means of its external mucus layer. This layer was comparatively thick and seemed to consist to a large extent of glycoprotein containing neutral polysaccharide. It lacked excitatory substances that were present in the mucus of related non-symbiotic fishes; it did not appear to contain specific substances that either chemically masked stimulatory compounds or else inhibited nematocyst discharge. The mucus layer thus provided a thick inert barrier between *A. clarkii*'s body and the anemone's nematocytes. It was clear that the clownfish could become covered in a small amount of anemone mucus as a result of contact with its host; this was perhaps beneficial in that it increased the thickness of the inert layer covering the fish, but there was no evidence that the clownfish was dependent upon the anemone mucus for protection. Protection of *A. clarkii* from nematocyst discharge appeared to be primarily the result of the mucus layer which it produced.

DISCUSSION

The experiments on *A. clarkii* and *S. haddoni* indicate that the symbiosis between the two species can be regarded as a problem of cellular recognition. The discharge of the anemone's stinging cells is preceded by a complex recognition process (Lubbock, 1979*a*), probably involving distinctive organelles associated with the exterior surface of the cells (Mariscal & Bigger, 1976; Lubbock & Shelton, 1981). The stinging response occurs only on contact with a substrate of suitable chemical composition (for example, prey) and cannot be elicited by mechanical stimulation alone (Lubbock, 1979*a*). *A. clarkii* produces a thick external mucus layer which does not appear to be recognized as foreign by the receptors associated with nematocyte discharge. The clownfish's mucus lacks excitatory substances that are present in the mucus of non-symbiotic species and thus provides an inert barrier between the nematocyte receptors and the fish's flesh (the latter is readily stung).

It is tempting to conclude that a similar mechanism is found in associations between other species of clownfish and anemone, if not between many cnidarian symbionts and their hosts. However, many earlier workers used different species of clownfish and/or anemone to those above, and it is possible that certain of the differences between my results and those of previous studies may be attributable to interspecific variations in the protective mechanism. Nevertheless, the concept of achieving protection from nematocyst discharge by means of an external layer of inert substances would seem to form a useful starting hypothesis in investigations of other cnidarian symbioses.

As previously indicated, protection of *A. clarkii* from *S. haddoni* was not found to be primarily dependent upon a coating of anemone mucus. However, the extent to which a layer of anemone mucus can provide a foreign object with protection from stinging remains a problem of interest; it is presumably related to the thickness of the layer and to the excitatory abilities of the underlying substrate. In the above experiments the mucus of non-symbiotic fishes, which is highly excitatory, could not be protected by coating with a thin layer of anemone mucus. Perhaps, however, a substrate that is only slightly stimulatory could be masked by such a coating.

The clownfish/anemone association has led to a variety of distinctive behaviour patterns on the part of the fish that are not found in non-symbiotic species; certain of these can be related to the fishes' protection mechanism. For example, it appears that the ability of many compounds to incite nematocyst discharge is increased when accompanied by strong mechanical stimulation (Lubbock, 1979*a*). This suggests that the slow, wary manner in which *Amphiprion* initially make contact with an unfamiliar anemone (Schlichter, 1968; Mariscal, 1971; Fukui, 1973) may not only prevent the possibility of the fish being stung over a wide area (as might happen if it immediately dashed in among the tentacles) but may also reduce the level of stimulation provided by potentially excitatory substances on its body surface.

The process of acclimatization has received much attention in the literature (see Mariscal, 1970*a*). It is the process by which a clownfish that has been isolated from an anemone gradually increases contact with the anemone, while the latter's stinging response towards the fish decreases. In the *A. clarkii/S. haddoni* symbiosis this phenomenon is not marked: within 1 or 2 min of being introduced to *S. haddoni*, previously isolated fishes seem to be able to swim unharmed among the tentacles. Experiments on the mucus of *A. clarkii* indicate that it remains inert to *S. haddoni*'s nematocytes even after the fish has been separated from an anemone for several months (see above). On the other hand, it was found (Lubbock, unpublished observations) that *A. clarkii* was unable to enter the anemone *Gyrostoma hertwigi* without undergoing an acclimatization period of approximately 4 days. This was regardless of whether the fishes had previously been isolated or inhabiting *S. haddoni*, indicating that anemone substances derived from a previous host were not involved (cf. Lubbock, 1979*b*). It seems likely that *G. hertwigi*'s nematocytes detect substances present in the mucus of these fish which are insufficient in either quality or quantity to induce nematocyte discharge in *S. haddoni*. As shown in Lubbock (1979*a*), the responses of *G. hertwigi* to different organic compounds are not identical to those of *S. haddoni*. This raises the interesting question of what occurs during the process of acclimatization. In the light of my results on the *A. clarkii/S. haddoni* symbiosis, the most likely explanation for the acclimatization of *A. clarkii* to *G. hertwigi* would seem to be that the fishes are reducing the levels of certain stimulatory substances in their mucus, perhaps as a result of being stung by the anemone.

As indicated above, *A. clarkii* lacks excitatory (to *S. haddoni*) substances which are present in the mucus of related fishes that do not inhabit anemones. At present the nature of these substances is not clear. Circumstantial evidence, however, indicates that mucus or epidermal substances associated with disease prevention may be present in smaller amounts in clownfish than in other fishes not symbiotic with anemones. R. D. Sankey, at present the main U.K. importer of tropical marine fishes, has informed me (personal communication) that clownfishes, especially *A. clarkii*, are much more susceptible to skin infections (when maintained in aquaria without anemones) than other members of the family Pomacentridae; Mariscal (1970*c*) reported similar findings. In this respect it is interesting to note that the antibiotic enzyme lysozyme has been found in fish mucus (Fletcher & Grant, 1968; Murray & Fletcher, 1976); since lysozyme can induce nematocyte discharge in *S. haddoni* (Lubbock, 1979*a*), one would imagine that this enzyme would either be absent or else in reduced quantity in *A. clarkii* mucus. Presumably, under natural conditions the settlement of fish parasites on clownfish is discouraged by the tendency of the fishes to wipe their bodies against the anemone and also by the presence of small amounts of anemone mucus on the clownfish's body surface (see above).

An interesting question is how the absence of excitatory substances in *A. clarkii* mucus might have evolved. Presumably the ancestors of these fishes possessed excitatory substances in their mucus; if so, how did the clownfishes adapt themselves to living in anemones? The comparative ecology of species in the pomacentrid genus *Dascyllus* indicates a possible intermediate stage in the development of the symbiosis with sea anemones. Of the 9 known species of *Dascyllus*, 2 (*D. trimaculatus* and *D. albisella*) are found either among corals or in sea anemones; the remainder (excluding the poorly known *D. strasburgi*) are generally found in live coral heads and have never been seen in sea anemones (Allen, 1975; Randall & Allen, 1977). Corals are cnidarians and possess nematocytes; the power of the stinging response in many species of coral, however, appears to be significantly less than that of host anemones used by fishes. It is possible, therefore, that the ancestors of *A. clarkii* inhabited corals, developed protection from the corals' nematocytes and subsequently became able to enter sea anemones; in such a case inhabiting living corals might to a certain extent have served the fishes as a preadaptation to living in sea anemones.

The clownfish/anemone symbiosis may be regarded as visible evidence of a cellular recognition system that has been circumvented by an unrelated animal. Nematocytes are cells with a well-developed ability for distinguishing between animal substances that are self and non-self. While they do not respond to the glycoproteins and other substances on the surface of their host anemone, they are relatively easily excited by foreign proteins and glycoproteins. The clownfish, however, has managed to produce a mucus layer that is not recognized as foreign by the anemone's nematocytes.

REFERENCES

Abel, E. F. (1960). Liaison facultative d'un poisson (*Gobius bucchichii* Steindachner) et d'une anémone (*Anemonia sulcata* Penn.) en Mediterranée. *Vie et Milieu* **11**, 515–31.

Allen, G. R. (1972). *The Anemonefishes. Their Classification and Biology.* New Jersey: T.F.H. Publications.

Allen, G. R. (1975). *Damselfishes of the South Seas.* New Jersey: T.F.H. Publications.

Blösch, M. (1961). Was ist die Grundlage der Korallenfischsymbiose: Schutzstoff oder Schutzverhalten? *Naturwissenschaften* **48** (9), 387.

Blösch, M. (1965). Untersuchungen über das Zusammenleben von Korallenfischen *Amphiprion* mit Seeanemonen. Inaugural Dissertation, Eberhard-Karls-Univerität zu Tübingen.

Briggs, R. P. (1976). Biology of *Paranthessius anemoniae* in association with anemone hosts. *Journal of the Marine Biological Association of the United Kingdom* **56**, 917–24.

Bruce, A. J. (1976). Shrimps and prawns of coral reefs, with special reference to commensalism. In *Biology and Geology of Coral Reefs* (ed. O. A. Jones and R. Endean), **3** (2). London: Academic Press.

Buhk, F. (1939). Lebensgemeinschaft zwischen Riesenseerose und Korallenfischen. *Wochenschrift für Aquarien und Terrarienkunde* **46,** 672–4.

Caspers, H. (1939). Histologische Untersuchungen über die Symbiose zwischen Aktinien und Korallenfischen. *Zoologischer Anzeiger* **126,** 245–53.

Collingwood, C. (1868). Note on the existence of gigantic sea-anemones in the China Sea, containing within them quasi-parasitic fish. *Annals and Magazine of Natural History* **4** (1), 31–3.

Conklin, E. J. & Mariscal, R. N. (1977). Feeding behavior, ceras structure and nematocyst storage in the aeolid nudibranch *Spurilla neapolitana* (Mollusca). *Bulletin of Marine Science* **27,** 658–67.

Dahl, E. (1961). The association between young whiting, *Gadus merlangus* and the jelly fish *Cyanea capillata. Sarsia* **3,** 47–55.

Davenport, D. & Norris, K. S. (1958). Observations on the symbiosis of the sea anemone *Stoichactis* and the pomacentrid fish, *Amphiprion percula. Biological Bulletin* **115** (3), 397–410.

Day, R. M. & Harris, L. G. (1978). Selection and turnover of coelenterate nematocysts in some aeolid nudibranchs. *Veliger* **21,** 104–9.

Edmunds, M. (1966). Protective mechanisms in the Eolidacea (Mollusca: Nudibranchia). *Journal of the Linnean Society of London, Zoology* **46,** 27–71.

Eibl-Eibesfeldt, I. (1960). Beobachtungen und Versuche an Anemonenfischen (*Amphiprion*) der Malediven und der Nicobaren. *Zeitschrift für Tierpsychologie* **17** (1), 1–10.

Fishelson, L. (1965). Observations and experiments on the Red Sea anemones and their symbiotic fish *Amphiprion bicinctus. Bulletin of the Sea Fisheries Research Station, Haifa* **39,** 1–14.

Fletcher, T. C. & Grant, P. T. (1968). Glycoproteins in the external mucus secretions of the plaice, *Pleuronectes platessa*, and other fishes. *The Biochemical Journal* **106,** 12P.

Fricke, H. W. (1967). Garnelen als Kommensalen der tropischen Seeanemone *Discosoma. Natur und Museum* **97,** 53–8.

Fricke, H. W. (1974). Öko-Ethologie des monogamen Anemonenfisches *Amphiprion bicinctus. Zeitschrift für Tierpsychologie* **36,** 429–512.

Fukui, Y. (1973). Some experiments on the symbiotic association between sea anemone and *Amphiprion. Publications of the Seto Marine Biological Laboratory* **20,** 419–30.

Gohar, H. A. F. (1948). Commensalism between fish and anemone. *Publications of the Marine Biological Station, Ghardaqa* **6,** 35–44.

Gordon, I. (1960). Additional note on the porcellanid sea-anemone association. *Crustaceana* **1,** 166–7.

Graefe, G. (1963). Die Anemonen-Fisch-Symbiose und ihre Grundlage nach Freilanduntersuchungen bei Eilat/Rotes Meer. *Naturwissenschaften* **50** (11), 410.

Graefe, G. (1964). Zur Anemonen-Fisch-Symbiose, nach Freilanduntersuchungen bei Eilat/Rotes Meer. *Zeitschrift für Tierpsychologie* **21** (4), 468–85.

Graham, A. (1938). The structure and functions of the alimentary canal of aeolid molluscs, with a discussion of their nematocysts. *Transactions of the Royal Society of Edinburgh* **59,** 267–307.

Hackinger, A. (1959). Freilandbeobachtungen an Aktinien und Korallenfischen. *Mitteilungen der Biologischen Station Wilhelminenburg* **2,** 72–4.

Hanlon, R. T. & Kaufman, L. (1976). Associations of seven West Indian reef fishes with sea anemones. *Bulletin of Marine Science* **26** (2), 225–32.

Harris, L. G. (1971). Nudibranch associations as symbioses. In *Aspects of the Biology of Symbiosis* (ed. T. C. Cheng). London: Butterworths.

Hayes, D. A. & Grayson, R. F. (1968). The freshwater hydras: symbionts and predators. *Countryside* **21,** 23–8.

Horst, R. (1902). On a case of commensalism of a fish (*Ampriprion intermedius* Schleg.) and a large sea anemone (*Discosoma* sp.). *Notes of the Leyden Museum* **23,** 181–2.

Kepner, W. A. & Nuttycombe, J. W. (1929). Further studies on the nematocysts of *Microstomum caudatum. Biological Bulletin* **57,** 69–81.

KOENIG, O. (1960). Verhaltensuntersuchungen an Anemonenfischen. *Die Pyramide* **8** (2), 52–6.

LUBBOCK, R. (1979*a*). Chemical recognition and nematocyte excitation in a sea anemone. *Journal of Experimental Biology* **83**, 283–92.

LUBBOCK, R. (1979*b*). Mucus antigenicity in sea anemones and corals. *Hydrobiologia* **66**, 3–6.

LUBBOCK, R. (1980). Why are clownfishes not stung by sea anemones? *Proceedings of the Royal Society of London, B* **207**, 35–61.

LUBBOCK, R. & SHELTON, G. A. B. (1981). Electrical activity following cellular recognition of self and non-self in a sea anemone. *Nature, London* (in the Press).

MANSUETI, R. (1963). Symbiotic behaviour between small fishes and jellyfishes, with new data on that between the Stromateid, *Peprilus alepidotus*, and the Scyphomedusa, *Chrysaora quinquecirrha*. *Copeia* **1963**, 40–80.

MARISCAL, R. N. (1965). Observations on acclimation behaviour in the symbiosis of anemone fish and sea anemones. *American Zoologist* **5**, 694.

MARISCAL, R. N. (1966). A field and experimental study of the symbiotic association of fishes and sea anemones. Ph.D. thesis, University of California, Berkeley.

MARISCAL, R. N. (1967). A field and experimental study of the symbiotic association of fishes and sea anemones. *Dissertation Abstracts* **2**, 388.

MARISCAL, R. N. (1970*a*). An experimental analysis of the protection of *Amphiprion xanthurus* Cuv. and Val. and some other anemone fishes and sea anemones. *Journal of Experimental Marine Biology and Ecology* **4**, 134–49.

MARISCAL, R. N. (1970*b*). A field and laboratory study of the symbiotic behavior of fishes and sea anemones from the tropical Indo-Pacific. *University of California Publications in Zoology* **91**, 1–33.

MARISCAL, R. N. (1970*c*). The nature of the symbiosis between Indo-Pacific anemone fishes and sea anemones. *Marine Biology* **6** (1), 58–65.

MARISCAL, R. N. (1971). Experimental studies on the protection of anemone fishes from sea anemones. In *Aspects of the Biology of Symbiosis* (ed. T. C. Cheng). London: Butterworths.

MARISCAL, R. N. (1972). Behavior of symbiotic fishes and sea anemones. In *Behavior of Marine Animals 2* (ed. H. E. Winn and B. L. Olla). New York: Plenum Press.

MARISCAL, R. N. (1974). Nematocysts. In *Coelenterate Biology: Reviews and New Perspectives* (ed. L. Muscatine and H. M. Lenhoff). London: Academic Press.

MARISCAL, R. N. & BIGGER, C. H. (1976). A comparison of putative sensory receptors associated with nematocysts in an anthozoan and a scyphozoan. In *Coelenterate Ecology and Behavior* (ed. G. O. Mackie). New York: Plenum Press.

MARTIN, C. H. (1908). The nematocysts of turbellaria. *Quarterly Journal of the Microscopical Society* **52**, 261–77.

MOSER, J. (1931). Beobachtungen über die Symbiose von *Amphiprion percula* (Lacépède) mit Aktinien. *Sitzungsberichte der Gesellschaft naturforschender Freunde zu Berlin* **2**, 160–7.

MURRAY, C. K. & FLETCHER, T. C. (1976). The immunohistochemical localisation of lysozyme in plaice (*Pleuronectes platessa* L.) tissues. *Journal of Fish Biology* **9**, 329–34.

NAYAR, K. N. & MAHADEVAN, S. (1967). Underwater ecological observations in the Gulf of Mannar, off Tuticorin. 5. On sea anemones and the fishes *Amphiprion* and *Dascyllus* found with them. *Journal of the Marine Biological Association of India* **7**, 458–9.

NORMARSKI, G. (1955). Microinterféromètre differentiel à ondes polarisées. *Journal de Physique* **16**, 9 S.

PADAWER, J. (1968). The Nomarski interference-contrast microscope. An experimental basis for image interpretation. *Journal of the Royal Microscopical Society* **88**, 305–49.

RANDALL, H. A. & ALLEN, G. R. (1977). A revision of the damselfish genus *Dascyllus* (Pomacentridae) with the description of a new species. *Records of the Australian Museum* **31**, 349–85.

REES, W. J. (1967). A brief survey of the symbiotic associations of cnidaria with mollusca. *Proceedings of the Malacological Society of London* **37**, 213–31.

ROSS, D. M. (1974). Behavior patterns in associations and interactions with other animals. In *Coelenterate Biology: Reviews and New Perspectives* (ed. L. Muscatine and H. M. Lenhoff). London: Academic Press.

RÜTZLER, K. (1971). Bredin-Archbold-Smithsonian biological survey of Dominica: burrowing sponges, genus *Siphonodictyon* Bergquist, from the Caribbean. *Smithsonian Contributions to Zoology* **77**, 1–37.

SCHLICHTER, D. (1967). Zur Klärung der Anemonen-Fisch-Symbiose. *Naturwissenschaften* **54**, 569.

SCHLICHTER, D. (1968). Das Zusammenleben von Riffanemonen und Anemonenfischen. *Zeitschrift für Tierpsychologie* **25**, 933–54.

SCHLICHTER, D. (1970). *Thalassoma amblycephalus* ein neuer Anemonenfisch Typ. *Marine Biology* **7** (3), 269–72.

SCHLICHTER, D. (1972). Chemische Tarnung. Die stoffliche Grundlage der Anpassung von Anemonenfischen an Riffanemonen. *Marine Biology* **12**, 137–50.

SCHLICHTER, D. (1975). Produktion oder übernahme von Schutzstoffen als Ursache des Nesselschutzes von Anemonenfischen? *Journal of Experimental Marine Biology and Ecology* **20**, 49–61.

SCHLICHTER, D. (1976). Macromolecular mimicry: substances released by sea anemones and their role in the protection of anemone fishes. In *Coelenterate Ecology and Behavior* (ed. G. O. Mackie). New York: Plenum Press.

SLUITER, C. P. (1888). Ein merkwürdiger Fall von Mutualismus. *Zoologischer Anzeiger* **11**, 240–3.

THIEL, H. (1979). Assoziationen von Quallen und Fischen. *Natur und Museum* **109**, 356–60.

THOMPSON, T. E. & BENNETT, I. (1970). Observations on Australian Glaucidae (Mollusca: Opisthobranchia). *Zoological Journal of the Linnean Society* **49**, 187–97.

VERWEY, J. (1930). Coral reef studies. I. The symbiosis between damselfishes and sea anemones in Batavia Bay. *Treubia* **12**, 305–66.

WRIGHT, T. S. (1858). On the cnidae or thread-cells of the Eolidae. *Proceedings of the Royal Physical Society of Edinburgh* **1**.

Parasitology (1981), **82**, 335–342
With 1 figure in the text

Helminths and the transmammary route of infection

GROVER C. MILLER
Department of Zoology, North Carolina State University, Raleigh, N.C. 27650

(*Accepted* 4 *September* 1980)

SUMMARY

The transfer of infective larval stages of helminths via the mammary glands is probably more common than generally recognized. Recent investigations of both natural and experimental infections of various animals have shown that the transmammary transmission of some helminths is a major avenue of infection. Other studies indicate that there are at least 13 helminth parasites which may be transmitted as prenatal infections and at least 17 transmitted via the mammary glands. The majority of these are nematodes. However, in one case the tetrathyridia of the cestode *Mesocestoides* was observed. In this report the life-cycle and transmission of 2 species of diplostomatid trematodes in the genus *Pharyngostomoides* is described. The raccoon, *Procyon lotor*, is the only definitive host; a planorbid snail, *Menetus dilatatus*, and a branchiobdellid annelid, *Cambarincola osceola*, a commensal on crayfishes, are intermediate hosts. Records for a period of 11 years are now available for the maternal transmission and longevity of this trematode. During this time one infected female raccoon was maintained in the laboratory on a diet of commercial dog food, sweet potatoes and corn. She produced 25 offspring, most of which were infected with *Pharyngostomoides*, through the 6th litter. There was a declining number of worms in each litter, ranging from nearly 2000 in one of the 1st litter, to none in the 7th litter. After nearly 12 years she continues to pass a few trematode ova. Rather than assume she has retained the same adult worms for 12 years, it is reasonable to conclude that mesocercariae move through her body and eventually mature in the intestine. Because the mesocercariae have a predilection for the lactating mammary gland, it seems likely that a hormonal influence is present. The transmammary transmission of larvae is a viable alternative in the life-cycles of a number of helminths, and in some instances it is probably the major route of infection.

INTRODUCTION

Most helminth infections of animals in early life have been viewed as being probably prenatal and little evidence existed for the peri-parturient transmission, especially for trematodes. The transfer of infective-stage larvae, observed by Salzer (1916) with *Trichinella spiralis* infections of humans, remained unconsidered as a cause of neonatal helminth infection until Lyons & Olsen (1960)

Table 1. *Prenatal and transmammary transmission of helminth parasites*

	Prenatal		Milk		
Nematodes	Natural	Exper.	Natural	Exper.	Undetermined
(1) *Ancylostoma caninum*	×	×	×	×	.
(2) *Arteonema (Setaria) cervi*	×	.	.	.	.
(3) *Dictyocaulus filaria*	×	.	.	.	.
(4) *Dirofilaria immitis*	×	.	.	.	.
(5) *D. repens*	×	.	.	.	.
(6) *Muellerius capillaris*	.	.	.	.	×
(7) *Nippostrongylus brasiliensis*	.	.	.	×	.
(8) *Protostrongylus stilesi*	×	.	.	.	.
(9) *Stephanurus dentatus*	.	.	.	.	×
(10) *Strongyloides fuelleborni*	×	.	.	.	.
(11) *S. papillosus*	×	×	×	×	.
(12) *S. ransomi*	×	×	×	×	.
(13) *S. ratti*	.	.	×	×	.
(14) *S. westeri*	.	.	×	.	.
(15) *Toxocara canis*	×	×	.	×	.
(16) *T. cati*	.	.	×	×	.
(17) *T. (Neoascaris) vitulorum*	.	.	×	.	.
(18) *Trichinella spiralis*	.	.	×	×	.
(19) *Uncinaria lucasi*	.	.	×	×	.
(20) *U. stenocephala*	.	.	×	.	.
Trematodes					
(21) *Fasciola hepatica*	×	.	.	.	.
(22) *Pharyngostomoides adenocephala*	.	.	×	× (?)	.
(23) *P. procyonis*	.	.	×	×	.
(24) *Schistosoma japonicum*	×	×	× (?)	.	.
Cestodes					
(25) *Cysticerus bovis*	.	.	.	×	.
(26) *Echinococcus granulosus*	×	.	.	.	.
(27) *Mesocestoides corti*	.	.	×	×	.

showed that infection of the seal *Callorhinus ursinus L.* with the hookworm *Uncinaria lucasi* was accomplished by transcolostral passage.

Recent investigations with both experimental and natural infections of animals have shown that some neonatal infections of helminths, once considered to be prenatal, are either partially or exclusively milk-borne. Nine species of nematodes cited by Stone & Smith (1973) are transmitted by milk-borne larvae. Stoye (1976) reviewed both prenatal and milk-borne helminth infections in animals. Today, 13 helminth parasites have been shown capable of prenatal infection and 17 helminth parasites capable of transmammary infection. However, not all evidence presented by some authors is convincing. Table 1 lists those helminths using prenatal or transmammary transmission routes in their life-cycles. The exact routes of 2 of these (*Meullerius capillaris* and *Stephanurus dentatus*) have not been determined.

Although some of the earlier observations on the presence of larval helminths in the mammary glands was incidental (Salzer, 1916; Narabayashi, 1914), more recent investigations have shown that transmammary transmission may be a major part of the life-cycle. This is especially true of nematodes (Olsen & Lyons, 1965; Stone & Girardeau, 1967; Moncol & Grice, 1974). For example, both *Toxocara canis* larvae (Enigk & Stoye, 1967) and *T. cati* larvae (Swerczek, Nielsen, & Helmboldt, 1971) have been found in transmammary passage, although the prenatal route maybe more important. Conversely, with *Ancylostoma caninum* larvae (Stone, Peckham & Smith, 1970; Miller, 1970) and *Uncinaria lucasi* larvae (Olsen & Lyons, 1965) infection takes place largely or exclusively via the colostrum and milk. Other species, especially those in the genus *Strongyloides* (see Table 1) are known to utilize transmammary passage. The reports of Zamirdin & Wilson (1974), Wilson (1977), Wilson, Cameron & Scott, (1978*a*, *b*) and Wilson (1979) are interesting for their thorough work on the life-history of *Strongyloides ratti*. Wilson (1977) suggested that neuroendocrine-related changes accompanying the sucking stimulus of young rats may influence the presence of 3rd-stage larvae of *S. ratti* in the mammary gland. Wilson (1979) reported the use of certain isotopes as a possible tool for tracking the migrating larvae. We agree with Wilson (1977) that there are probably hormone operators which influence the concentration of larval helminths in the lactating mammary glands. The report of Brown & Girardeau (1977) of the recovery of larval *S. fuelleborni* from human milk illustrates a greater awareness of this method of infection and its influence on medical and veterinary parasitology.

The passage of tapeworms via the colostrum or milk is less well known. Without special safeguards it is difficult to demonstrate infection by milk-borne transmission. Some such suspected infections may, in reality, be prenatal in origin. Conversely, some common milk-transmitted infections may have been assumed to be prenatal in origin. The colostral passage of the cestode *Mesocestoides corti* has been demonstrated in an experimental infection of the mouse (Hess, 1972). Apparently this is the only report for milk-borne transmission of cestodes.

Stoye's (1976) review reveals that prenatal transfer of *Fasciola hepatica* has been observed in humans (Straub, 1958) and horses (Pozajic, 1914) and others have noted the same in cattle (Rees, Sykes & Richard, 1975). Early references have been made to infections in humans with *Schistosoma japonicum* following suckling (Fujinami & Nakamura, 1911; Narabayashi, 1914) and to prenatal infections in dogs, rabbits and guinea pigs following experimental infections (Narabayashi, 1914). The possibility of neonatal infection in humans with blood flukes should not be overlooked.

STUDIES ON NEONATAL TRANSMISSION OF TREMATODES

The only trematodes presently known to utilize transmammary passage are the diplostomes, *Pharyngostomoides procyonis* Harkema, 1942 and *Pharyngostomoides adenocephala* Beckerdite, Miller and Harkema, 1971. Both of these species are parasites of the intestine of the raccoon, *Procyon lotor* and are found only in this host animal throughout its range (Fig. 1). Thus, in order to obtain raccoons free of the parasite, it was necessary to collect them from an off-shore island such as

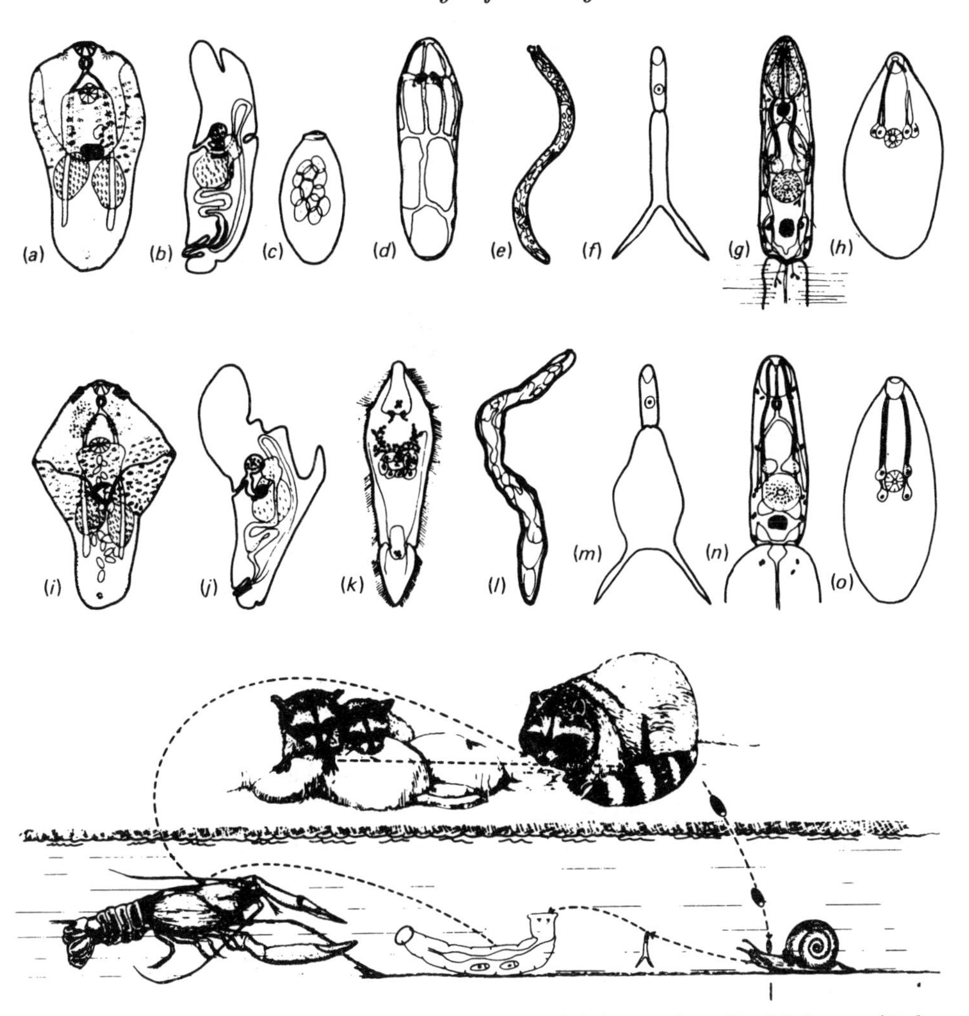

Fig. 1. *a–h*. *Pharyngostomoides procyonis*. (*a*) Adult – entire. (*b*) Adult – sagittal view. (*c*) Egg. (*d*) Miracidium showing epidermal plates. (*e*) Sporocyst. (*f*) Cercaria – entire. (*g*) Cercaria – body without tail. (*h*) Mesocercaria.

i–o. *Pharyngostomoides adenocephala*. (*i*) Adult – entire. (*j*) Adult – sagittal view. (*k*) Miracidium showing ciliated epidermis. (*l*) Sporocyst. (*m*) Cercaria – entire. (*n*) Cercaria – body without tail. (*o*) Mesocercaria.

Baldhead (Smith Island) on the North Carolina coast. Apparently the intermediate hosts for these parasites are absent from the islands. *P. procyonis* and *P. adenocephala* occur together in the raccoon and originally were thought to be variants of the same species until life-history studies proved otherwise. *P. adenocephala* is commonly found in the first 30 centimeters of the small intestine, whereas *P. procyonis* is generally beyond this area but overlapping to some extent. The larval stages vary considerably but *Menetus dilatatus* is the snail host for both trematodes (Beckerdite, Miller & Harkema, 1971).

Several investigators (Odlaug, 1940; Pearson, 1956; Johnson, 1968; Harris, Harkema & Miller, 1967*a*, 1970) have worked on the life-cycles of diplostomatid trematodes and in each case have identified some species of amphibian, either tadpoles and/or adult frogs and salamanders as a 2nd intermediate host. Certain reptilian, avian and mammalian hosts have been identified as paratenic hosts in other diplostomatids. None of these was involved in the life-cycle of either species of *Pharyngostomoides*. Numerous feeding experiments involving fishes, amphibians and reptiles proved unsuccessful. Likewise, many tadpoles and some frogs of various species were exposed to cercariae but none became infected.

The first evidence on the life-cycle was obtained when one raccoon became infected after eating crayfishes. A subsequent feeding of crayfishes to a second raccoon, however, proved negative and again, the intermediate host was questionable. Later, other crayfishes were carefully dissected and mesocercariae (Fig. 1*h* and *o*) were found inside the branchiobdellid annelids associated with the crayfish. Nearly 700 crayfishes were examined and 9921 annelids were recovered. In one instance, an annelid contained more than 300 mesocercariae but the usual number was 2–4/infected worm. Approximately 15% of the annelids were infected with 8703 small mesocercariae or with 489 larger ones. We believe that these two mesocercariae represent the two species in the genus *Pharyngostomoides*.

Infected annelids were fed to a variety of hosts including day-old chicks, mice, kittens and raccoons. Three chicks were fed annelids containing approximately 250 mesocercariae of both sizes. The hosts were necropsied at different times during the next 2 weeks and all organs were examined for any stage of the parasite. It was felt that the chick might serve as a paratenic host. The muscles were carefully checked for mesocercariae, but none was found. Likewise, 2 mice were fed infected annelids; both were necropsied 20 days later and all organs were negative. Having obtained worms with eggs from mature cats which had been fed infected raccoon lungs, it seemed possible that kittens might serve as an adequate definitive host. Such was not the case, as 3 kittens in 3 separate experiments were not infected.

The successful feeding experiments involved only the raccoon. This is not surprising as the raccoon is the only definitive host reported for these trematodes. From our experiments, it would appear that there are no paratenic hosts and the parasites matured in the intestine of the cat only after exposure to 'somatic migration' in the raccoon.

A raccoon, free of the parasite, was fed 140 annelids containing 265 mesocercariae and later 76 annelids containing 141 mesocercariae. Eggs first appeared in the faeces 30 days later and at necropsy 11 adults of *P. procyonis* were recovered. A second raccoon, from Smith Island, was fed infected annelids; eggs first appeared 35 days after the initial feeding. The animal was necropsied after 55 days and 18 adult worms were recovered. Two of these were not ovigerous. No worms were found in the lungs or other organs. In addition to the 2 raccoons mentioned above, 4 others were infected the same way and necropsied at intervals to check the route of the mesocercariae.

EVIDENCE FOR MATERNAL TRANSMISSION OF TREMATODES IN THE RACCOON

On 15 April 1961, a 9-week-old raccoon which had been born in captivity, died and was examined for parasites. Seventy-eight specimens of *P. procyonis* were recovered, but the mode of infection was not postulated (Harkema & Miller, 1964). Subsequent to this finding, more diligent efforts were made to examine those animals born in captivity and to guard their diets. Later observations were concentrated on young raccoons prior to weaning. In addition to the infected raccoon reported in 1961, there were 3 in 1964, 3 in 1969, 4 in 1970, 2 in 1971, 4 in 1972, 5 in 1973 and 3 (of 5) in 1974, all born in captivity and all infected with *Pharyngostomoides*. Two of the 5 born in 1974 were uninfected and neither of the 2 born in 1975 was infected, indicating a gradual exhaustion of mesocercariae in the mother.

Our preliminary report on maternal transmission (Harris, Harkema & Miller 1967*b*) indicated that several late-term foetus and neonatal raccoons from infected mothers were not infected, which seemed to rule out transplacental infection. Numerous mesocercariae were found in the mammary glands of lactating females and in the abdominal muscles, lungs and body cavities of raccoons. The mesocercariae required an extra-intestinal phase of maturation to become a metacercaria (more properly a diplostomulum) before moving to the intestine. This appears to be the case even in transmammary transmission. Necropsy of a 10-week-old raccoon, born in captivity, revealed the following: 11 *Pharyngostomoides* spp. from the stomach, 1753 from the small intestine; 5 mesocercariae and 14 metacercariae (diplostomula) from the lungs.

We now have records of longevity and maternal transmission extending over a period of 11 years. A male raccoon, trapped in May 1968, and maintained on a diet of commercial dog food, boiled sweet potatoes and corn, continued to pass eggs in the faeces until death (presumably from old age and pneumonia) in July 1979. Six viable *P. procyonis* adults were recovered from the intestine. A female, trapped in February 1969, and maintained on the same diet, also continues to pass eggs. This female was bred with the male mentioned above each spring for 7 years. She had, on average, nearly 4 young each year, with a total of 25 offspring. These young raccoons have been consistently infected with *P. procyonis* each year through the 6th litter. In the 6th litter, 3 of the 5 surviving young (a 6th was born dead) were infected, as determined by faecal examination and necropsy. However, there has been a declining number of worms in each litter, ranging from nearly 2000 in one raccoon of the 1st litter, to 2 or none in the 6th litter and none in the 7th litter. Again it is emphasized that none of these host animals had access to possible intermediate hosts of the trematode. These observations confirm the maternal transmission of this trematode and the probability that raccoons are infected for life.

We are indebted to Dr Perry Holt of Virginia Polytechnic Institute and State University for identifying the annelid second intermediate host as *Cambarincola osceola* Hoffman. We also thank Dr Horton H. Hobbs, Jr for identifying the 2 species of crayfishes which harboured the annelids. They are *Cambarus latimanus* (LeConte) and *Procambarus acutus acutus* (Girard).

REFERENCES

BECKERDITE, F. W., MILLER, G. C. & HARKEMA, R. (1971). Observations on the life cycle of *Pharyngostomoides* spp. and the description of *P. adenocephala* sp.n. (Strigeodea: Diplostomatidae) from the raccoon, *Procyn lotor* (L.) *Proceedings of the Helminthological Society of Washington* **38**, 149–56.

BROWN, R. C. & GIRARDEAU, M. H. F. (1977). Transmammary passage of *Strongyloides* sp. larvae in the human host. *American Journal of Tropical Medicine and Hygiene* **26**, 215–19.

ENIGK, K. & STOYE, M. (1967). Untersuchungen über den Infektionsweg von *Ancylostoma caninum* Ercolani 1859 (Ancylostomidae) beim Hund. Kongressbericht über die III Tagung der Deutschen tropenmedizinischen Gesellschaft e.v. Hamburg, vom 20 bis 22 April. München-Berlin-Wien: Urban Schwarzenberg.

FUJINAMI, A. & NAKAMURA, H. (1911). On the prophylaxis of schistosomiasis and some investigations on infection with this disease. *Chugai Iji Shinpo* **753**, 1009–27.

HARKEMA, R. & MILLER, G. C. (1964). Helminth parasites of the raccoon, *Procyon lotor* in the Southeastern United States. *Journal of Parasitology* **50**, 60–6.

HARRIS, A. H., HARKEMA, R. & MILLER, G. C. (1967*a*). Life history and taxonomy of *Diplostomum variable* (Chandler, 1932) Trematoda: Diplostamatidae). *Journal of Parasitology* **53**, 577–83.

HARRIS, A. H., HARKEMA, R. & MILLER, G. C. (1967*b*). Maternal transmission of *Pharyngostomoides procyonis* Harkema, 1942 (Trematoda: Diplostomatidae). *Journal of Parasitology* **53**, 1114–15.

HARRIS, A. H., HARKEMA, R. & MILLER, G. C. (1970). Life cycle of *Procyotrema marsupiformis* Harkema and Miller, 1959. (Trematoda: Strigeoidea: Diplostomatidea). *Journal of Parasitology* **56**, 297–301.

HESS, E. (1972). Transmission maternelle de Tetrathyridia (Mesocestoides, Cyclophyllidea) chez la souris blanche. *Comptes rendus, Académie des sciences, Paris* **274**, 596–9.

JOHNSON, A. D. (1968). Life history of *Alaria marcianae* (LaRue, 1917) Walton, 1949 (Trematoda: Diplostomatidae). *Journal of Parasitology* **54**, 324–32.

LYONS, E. T. & OLSEN, O. W. (1960). Report on the seventh summer of investigations on hookworms, *Uncinaria lucasi* Stiles, 1901, and hookworm diseases of fur seals, *Callorhinus ursinus* Linn. on the Pribilof Islands, Alaska from 15 June to 3 October 1960. *United States Department of Interior, Fish and Wildlife Service.*, Washington, D.C. p. 26 Hectograph.

MILLER, T. A. (1970). Prenatal-colostral infection of pups with *Ancylostoma caninum*. *Journal of Parasitology* **56** (4), 239.

MONCOL, D. J. & GRICE, M. J. (1974). Transmammary passage of *Strongyloides papillosus* in the goat and sheep. *Proceedings of the Helminthological Society of Washington* **41**, 1–3.

NARABAYASHI, H. (1914). Beiträge zur Frage der kongenitalen Invasion von *Schistosomum japonicum*. *Verhandlungen der Japanischen pathologischen Gesellschaft* **4**, 123.

ODLAUG, T. Q. (1940). Morphology and life history of the trematode, *Alaria intermedia*. *Transactions of the American Microscopical Society* **59**, 490–510.

OLSEN, O. W. & LYONS, E. T. (1965). Life cycle of the hookworm, *Uncinaria lucasi* Stiles, 1901 (Nematoda: Ancylostomatidae) of fur seals, *Callorhinus ursinus* Linn. on the Pribilof Islands, Alaska. *Journal of Parasitology* **51**, 689–700.

PEARSON, J. C. (1956). Studies on the life cycle and morphology of the larval stages of *Alaria arisaemoides* Augustine and Uribe, 1926 and *Alaria canis* LaRue and Fallis, 1936 (Trematoda: Diplostomatidae). *Canadian Journal of Zoology* **34**, 295–387.

POZAJIC, D. (1914). (Zit. bei Enigk, K. & Duwel, D. (1959)) in: Zur Haufigkeit der pranatalen Infektion mit *Fasciola hepatica* beim Rinde.) *Berliner und Münchener tierärztliche Wochenschrift* **72**, 362–3.

REES, J. B., SYKES, W. E. & RICHARD, M. D. (1975). Prenatal infection with *Fasciola hepatica* in calves. *Australian Veterinary Journal* **51**, 497–9.

SALZER, B. F. (1916). A study of an epidemic of fourteen cases of trichinosis with cures by serum therapy. *Journal of the American Medical Association* **67**, 579–80.

STONE, W. M. & GIRARDEAU, M. A. (1967). Transmammary passage of infective stage nematode larvae. *Veterinary Medicine/Small Animal Clinic* **62**, 252–3.

STONE, W. M., PECKHAM, J. C. & SMITH, F. W. (1970). Prenatal infections of *Ancylostoma caninum* among pups from bitches naturally infected. *Journal of the American Veterinary Medical Association* **157**, 2104–5.

STONE, W. M. & SMITH, F. W. (1973). Infection of mammalian hosts by milk-borne nematode larvae: a review. *Experimental Parasitology* **34**, 306–12.

STOYE, M. (1976). Ubersichtsreferat: Prantalae und galaklogene Helmintheninfektionen bei Haustieren. *Deutsche Tierärztliche Wochenschrift* **83**, 569–76.

STRAUB, G. (1958). Ein familiares Vorkommen der Fasciolose. Vortragsmateriel der *International Parasitologenkongresse* Budapest.

SWERCZEK, T. W., NIELSEN, S. W. & HELMBOLDT, C. F. (1971). Transmammary passage of *Toxocara cati* in the cat. *American Journal of Veterinary Research* **32**, 89–92.

WILSON, P. A. G. (1977). The effect of the suckling stimulus on the migration of *Strongyloides ratti* in lactating rats. *Parasitology* **75**, 233–9.

WILSON, P. A. G. (1979). Tracking radioactive larvae of *Strongyloides ratti* in the host. *Parasitology* **79**, 29–38.

WILSON, P. A. G., CAMERON, M. & SCOTT, D. S. (1978*a*). *Strongyloides ratti* in virgin female rats: studies of oestrous cycle effects and general variability. *Parasitology* **76**, 221–7.

WILSON, P. A. G., CAMERON, M. & SCOTT, D. S. (1978*b*). Patterns of milk transmission of *Strongyloides ratti*. *Parasitology* **77**, 87–96.

ZAMIRDIN, M. & WILSON, P. A. G. (1974). *Strongyloides ratti*: relative importance of maternal sources of infection. *Parasitology* **69**, 445–53.

Parasitology (1981), **82**, 489–509
With 2 plates

Parasites and the fossil record

S. CONWAY MORRIS
Department of Earth Sciences, The Open University, Milton Keynes MK7 6AA

(*Accepted* 29 *October* 1980)

INTRODUCTION

Parasitologists are fond of alluding to the possible antiquity of their chosen group. They may even place the origin in a specific period of geological time and trace its co-evolution, or lack thereof, with the hosts. These speculations usually make only passing comments on the actual fossil record, but a critical examination suggests that this area can throw much light on both the origins and evolution of parasitism. This review is concerned primarily with the fossil record and its immediate implications. It does not appraise parasite evolution in the light of evidence from sources, often of excellent quality, such as comparative anatomy and physiology or ecology. As some parasitologists may not be entirely familiar with the geological column and time scale, Table 1 depicts these and indicates some records of parasitism mentioned here, together with the major events in this history of life.

There are several approaches which may prove profitable. (1) The most obvious step is to look at fossil parasites or failing that, the reactions they provoked in the hard skeletal tissue of the host which is normally the only part of the organism to survive. These reactions, such as galls and tumours, fall into the category of trace fossils and have the disadvantage that unrelated parasites may produce similar disturbances. (2) Analysis of other better known fossil associations, such as commensalism, may show the evolution of an association through time and provide analogies to typical parasitic features like host specificity. (3) The evolution of many host groups is known in fair detail. Such data may indicate at least the earliest possible time of infestation and ideally allow the plotting of host–parasite co-evolution. (4) One of the greatest challenges for parasitology lies in understanding the macro-evolution of parasites. The fossil record is a fertile source of information on macro-evolution (Stanley, 1979). New evolutionary models, that are already revolutionizing traditional ways of looking at the fossil record, may prove applicable to understanding the origin and evolution of parasites. (5) The final avenue is through the widely accepted theory of plate tectonics. The almost continuous creation of oceans and movements of continents will promote, of course, effects such as isolation and migration of hosts and their parasites. It appears, however, that the biogeography of parasites can only be related in a very generalized manner to the dispersal and rejoining of continents. Ferris, Goseco & Ferris (1976) have attempted to cast the distribution of free-living nematodes into a framework of plate tectonics. Parasites also should be amenable to this type of study, but as yet plate tectonics has only provided a very broad introduction to some aspects of parasite evolution.

Table 1. *The geological column and records of fossil parasites in relation to some of the major events in the history of life*

Era	Period	Age (base of period, Millions of years)	Parasite event	Other event
Cainozoic	Pleistocene	1·8	Nematodes in frozen mammals	First *Homo*
	Pliocene	5		First hominids
	Miocene	22·5	Fossil turbellaria	
	Oligocene	38	Parasitic nematodes	
	Eocene	54	Parasitic nematodes and gordians	Major radiation of mammals
	Palaeocene	65		
Mesozoic	Cretaceous	141	Earliest undoubted parasitic copepod	Mass extinction
			Earliest parasitic insect	First angiosperms, placental mammals, and modern teleosts
			Earliest parasitic barnacles	
			?Earliest parasitic gastropods	
	Jurassic	195	Major radiation of parasitic isopods	First birds
			First parasitic isopods	
			?Earliest parasitic copepods	
	Triassic	232	First molluscan pearls	First mammals First dinosaurs
Palaeozoic	Permian	280	Earliest insect galls	Mass extinction
	Carboniferous	345	First nematodes Early myzostomids	First reptiles First winged insects
	Devonian	395	?Earliest myzostomids	First gymnosperms First amphibians
			Parasitic fungi	First insects (collembolans)
	Silurian	435	?Parasitic blisters on graptolites	First land plants First jawed fish
	Ordovician	500	Parasitic 'tubothecae' and blisters on graptolites	Major radiation of shelly invertebrates
			Parasitic galls in trilobites	
			First *Phosphannulus* in crinoids	
			Beginning of platy ceratid gastropod–crinoid association	
	Cambrian	570	?Pre-acanthocephalan (*Ancalagon*)	First fish (jawless)
			Parasitic traces in trilobites	Appearance of hard parts: first radiation of invertebrates
Precambrian	Vendian	680	?Earliest platyhelminths	Origin and radiation of metazoan phyla
			First inter-specific association in metazoans	

FOSSIL PARASITES

(1) *Introduction*

It is desirable to review briefly the extent of present knowledge on fossil parasites, which occurs in a widely scattered and often overlooked literature. It is notable that only a relatively small fraction of the parasites which one might expect to be preserved in the fossil record has actually been found. Furthermore, recognition of a fossil association as being parasitic may not be easy. An example of the problem of giving an unequivocal definition is evident in *Diorygma atrypophilia* (MacKinnon & Biernat, 1970) and *Burrinjuckia spiriferidophilia* (Chatterton, 1975). They are organisms of uncertain affinity, although MacKinnon & Biernat (1970) suggested a possible relationship with the phoronids. *D. atrypophilia* and *B. spiriferidophilia* inhabited tubes which are invariably located within the interior of ventral and dorsal valves respectively of Devonian brachiopods. They are orientated so as to be adjacent to the lophophore and to intercept feeding currents. Trophically these tubicolous organisms differ little from the commensal epizoans found clustered around the inhalent openings on the exterior surfaces of many brachiopods. It is clear, however, that the larvae of these organisms on entering the brachiopod feeding chamber, avoided ingestion, settled and penetrated the soft tissues. Having reached the underlying calcareous shell the invader provoked secretory cells of the brachiopod epithelium to build the elongate calcareous tube which they inhabited (MacKinnon & Biernat, 1970; Chatterton, 1975). The intricacy of this association with the precision of epizoan control over the secretion of the shell material indicates the difficulty experienced in classifying a number of fossil associations.

(2) *Nematodes and gordians*

As parasites are usually either soft-bodied or only lightly skeletonized their chances of fossilization are minimal. Surprisingly, the nematodes have a significant record, and this may be a reflection of a thick and relatively decay-resistant cuticle. Fossils of both free-living and parasitic nematodes are known and their occurrence is reviewed by Taylor (1935), Dollfus (1950) and Poinar (1977). Trace fossils in the form of sinusoidal markings from Eocene lacustrine sediments (Moussa, 1969, 1970) and borings in Upper Cretaceous foraminifers (Sliter, 1971) are ascribed to the activities of free-living nematodes. The possibility remains that the species which made these traces, or indeed actual fossil nematodes described as free-living (Pierce, 1964), were parasitic during other periods of their life-cycle.

Several examples of parasitic nematodes are recorded from the Cainozoic. *Heydenius antiquus*, known from a single specimen projecting from the anus of cerambycid beetle (*Hesthesis immortua*), comes from the Eocene lignite of the Rhine (Heyden, 1860, 1862). Three specimens of another species, *H. matutinus*, have been described in close association with a dipteran in Baltic amber of Oligocene age (Menge, 1866). A third case is an indeterminate larval nematode encysted in the striated abdominal muscles of a beetle (? *Chlorodema*) from the Eocene Geiseltal lignite of East Germany (Voigt, 1957). Parasitic nematodes have also been described from Pleistocene sousliks (Dubinin, 1948), woolly mammoths (Vereschagin, 1975) and a horse (Dubinina, 1972) found frozen in the permafrost of Siberia. The mould of a nematode has even been noted in a kidney-stone from a Pleistocene cave-bear (Tasnadi-Kubacska, 1962).

The potential for further work on specimens recovered from permafrost is obvious, although their comparative recency in terms of the geological time-scale suggests that new results will not be revolutionary. The faunas of the Baltic amber and Geiseltal lignite seem to be fairly well known. Further advances may be made by searching other deposits such as the Mexican Oligocene ambers, which have already yielded free-living nematodes (Poinar, 1977), and the older Cretaceous ambers. Voigt's (1957) discovery suggests that a reconnaissance programme of sectioning other fossilized muscle tissue could reveal additional encysted parasites. Dean (1902), for instance, described magnificently preserved shark muscle from the Devonian of Cleveland, USA in which the striations were clearly visible. Phosphatized fish muscles from the Solnhofen Limestone (Upper Jurassic) of Germany might be another productive source.

The earliest known nematodes are from the Lower Carboniferous (Calciferous Sandstone) of Scotland (Størmer, 1963) and Upper Carboniferous (Mazon Creek nodules) of Illinois (Schram, 1973). These nematodes were free-living and aquatic (?non-marine). The older specimens, described as *Scorpiophagus baculiformis* and *S. latus* are very interesting because they occur between the cuticle layers of a large scorpion and were evidently saprobiotic (Størmer, 1963). Strong arguments exist for the adoption of parasitism in 'phasmidian' nematodes via pre-adapted saprobiotic forms (Osche, 1963; Sudhaus, 1974). Parasitism in 'phasmidians', which form the bulk of parasitic forms, and 'aphasmidians' appear to have arisen repeatedly from terrestrial nematode stocks (Osche, 1963; Inglis, 1965). This Carboniferous scorpion may have been able to walk on land, but was evidently primarily aquatic (Størmer, 1963). This suggests that *Scorpiophagus* was not directly involved in the evolution of nematode parasitism. Further information could come from other cuticle preparations of Palaeozoic arthropods. Such preparations are already available for magnificently preserved chelicerates (eurypterids) from the Silurian of Estonia. No saprobiotic nematodes appear to have been found, although in passing it should be noted that curious tubular structures found in one such eurypterid may represent parasites (Clarke, 1921) and they deserve further study.

The gordian worms, which may be related to nematodes, are known from an incomplete fossil of *Gordius* from the Eocene Geiseltal lignite (Voigt, 1938). *Gordius* is regarded as a relatively advanced genus and this occurrence suggests that gordians had a considerable, but undefined, prior history (Sciacchitano, 1955).

(3) *Platyhelminths*

With the exception of some eggs and larvae recorded from archaeological sites (Gooch, 1975) the fossil record of parasitic platyhelminths is non-existent. Their frequent discovery in faecal material from archaeological excavations, together with similar remains of nematodes and occasionally acanthocephala, suggests that a detailed examination of fossil faeces (coprolites) may reveal eggs and other traces of parasites. Coprolites are abundant in many strata and their original organic rich composition may ensure very fine preservation of ingested parts. Some coprolites are attributable to specific groups of animals so that, if and when parasitic remains are discovered, inferences on their life-cycle may be possible.

With regards to the ancestry of platyhelminths there are a few clues. Fossils claimed to be turbellarians, presumably free-living, have been described from the Miocene of California (Pierce, 1960, 1964). More significant is the allocation by M. A. Fedonkin of the late Precambrian (Vendian) worms *Dickinsonia* and *Paleoplatoda* to the free-living platyhelminths (Palij, Posti & Fedonkin, 1979). These meagre hints may be supplemented by more circuitous approaches. Pearls in recent molluscs arise from a variety of irritants, but in many instances the sources are trematode larvae. The earliest molluscan pearls appear to be Triassic (Brown, 1940; Tasnadi-Kubacska, 1962), although molluscs date back to the early Cambrian. Whether this lack of molluscan pearls indicates a genuine absence of marine parasitic platyhelminths such as trematodes in the Palaeozoic is uncertain. Pearl-like objects have been recorded in association with conodonts of Cambrian to Carboniferous age and may have been provoked by parasites (Glenister, Klapper & Chauff, 1976). Conodonts are an enigmatic group of phosphatic microfossils that evidently formed the feeding apparatus of an unknown group. Encapsulated foreign bodies have also been reported within the conical shells of Silurian tentaculites, a group of uncertain affinities; but in this case a pearl-like structure is not apparent (Larsson, 1979).

Another line of enquiry which may prove profitable stems from the observation that shell growth in some modern gastropods is abnormal when the snail is infected with trematodes (Rothschild, 1936; Pan, 1963; Sturrock & Sturrock, 1971). Abnormalities are known in fossil gastropods (Tasnadi-Kubacska, 1962). None appear to have been linked with parasitism, but a critical search is clearly called for.

(4) *Acanthocephala*

This group is especially interesting both because of its extensive adaptation to parasitism and the problem of identifying its nearest relatives. The present consensus appears to favour a position for the acanthocephala near or within the ragbag of phyla which comprise the aschelminths, most notably the rotifers (Whitfield, 1971*b*; Graeber & Storch, 1978), and the priapulids (Meyer, 1933; Golvan, 1958). The discussion by Golvan (1958) featured the re-equipping of an acanthocephalan with the organs necessary for a free-living existence. The striking resemblance this hypothetical proto-acanthocephalan has to priapulids has been strengthened by the description of some fossil priapulids from the Middle Cambrian Burgess Shale of Canada (Conway Morris, 1977). The most abundant of the Burgess Shale priapulids, *Ottoia prolifica*, has already been compared with the acanthocephala (Meyer, 1933; Crompton, 1975), but a rarer priapulid known as *Ancalagon minor* (Pl. 1 D) has an even greater resemblance to the hypothetical proto-acanthocephalan. *Ancalagon* itself is too large (average adult length 6 cm) to be the immediate ancestor of the acanthocephala but a progenetic, or sexually mature juvenile, relative which would have belonged to the meiofauna (Nicholas, 1971) would be well placed to become parasitic. Like modern priapulids those in the Burgess Shale were infaunal and evidently active burrowers in the fine silts and muds. Nicholas (1971) and Whitfield (1971*a*) have both proposed that the ancestral free-living acanthocephala were burrowers. The proboscis hooks were used as anchors in marine sediments and were thereby pre-adapted for parasitism (Whitfield, 1971*a*). The Burgess Shale fauna contains numerous arthropods and their adoption as hosts by acanthocephala could date from about this time. If this did happen the incorporation of vertebrates into the life-cycle to form ultimately the primary hosts presumably occurred later (Van Cleave, 1947). The earliest vertebrates are Upper Cambrian fish (Repetski, 1977) but the acanthocephala may have invaded vertebrates at a later stage.

(5) *Arthropods*

A number of parasitic insects are known as fossils from Cainozoic deposits such as the Baltic ambers (Moodie, 1923; Tasnadi-Kubacska, 1962). The discovery of fossil mosquitoes and tsetse flies dating back to the early Cainozoic has led repeatedly to the suggestion that they were vectors of protozoan parasites. In passing it should be noted that the origin of vector systems could extend far back into the Palaeozoic. The occurrence in the guts of the Middle Cambrian priapulids (Conway Morris, 1977) of mollusc-like shells that were eaten alive, for instance, shows that the potential for vector systems existed more than 500 Ma ago.

Pre-Cainozoic parasitic insects seem to be much rarer. Riek (1970) described 2 distinctly different types of flea from the Lower Cretaceous of Australia. He noted that the fleas implied the presence of sparsely haired, warm-blooded animals, presumably marsupials. If this inference is correct then, as Riek notes, this alters profoundly present notions on the earliest evolution of marsupials. Ponomarenko (1976) noted a more unusual ?flea from the Lower Cretaceous of Transbaikalia. Again, a host could not be positively identified, but one possibility to be suggested was flying reptiles which may have been covered with fur.

Galls in fossil plants are also known. Most are probably due to parasitic insects (Berry, cited in Moodie, (1923)). The great majority are Cainozoic, but a few Mesozoic examples are known (Lesquereux, 1892; Alvin, Barnard, Harris, Hughes, Wagner & Wesley, 1967) and the earliest insect galls appear to be Permian (Potonie, 1893). Evidence for arthropod attack on Devonian (Kevan, Chaloner & Savile, 1975) and the Carboniferous plants (Scott, 1977) may include examples that could be classed as parasitism.

Fossils of marine parasitic arthropods are comparatively well known. Well preserved parasitic copepods were described from the gill chambers of a Lower Cretaceous fish from Brazil (Cressey & Patterson, 1973). These copepods were siphonostomes and significantly appear to be intermediate between those presently infesting fish and invertebrates. The copepods have been secondarily replaced by calcium phosphate and were found when the fish was prepared in weak acetic acid. The recovery from acid-digested residues of similarly preserved ostracodes dating as far back as Cambrian (Müller, 1979) shows that this technique may have an important application in elucidating the history of parasitic copepods. More tentative evidence for parasitic copepods comes from galls in Lower Jurassic (Lias) crinoid stems (Tetry, 1936), and Upper Jurassic (Callovian) (Mercier, 1936; Solov'yev, 1961) and Miocene echinoids (Margara, 1946).

Amongst the barnacles, several Cretaceous trace fossils have been assigned to the parasitic ascothoracicans. The evidence for this includes borings in echinoids (Madsen & Wolff, 1965) and cysts and excavations in octocoral spicules (Voigt, 1959, 1967). Somewhat surprisingly the more famous parasitic rhizocephalan barnacles appear to lack a fossil record.

The activities of parasitic isopods (Epicaridea) have been inferred from the characteristic swellings in the branchial region of decapod carapaces (Bachmayer, 1948; Housa, 1963; Förster, 1969; Radwanski, 1972). In no case, however, has the lightly skeletonized isopod been found beneath the carapace. Thus, while the swellings are regarded as being diagnostic of isopods they do not enable identification of specific families, although they are usually ascribed to members of the Bopyridae (Bachmayer, 1955; Housa, 1963; Förster, 1969). The earliest cases of infestation are from the Upper Jurassic (Oxfordian and Kimmeridgian), but in the uppermost Jurassic (Tithonian) the number of infested species had approximately doubled to about 10 (Förster, 1969; Radwanski, 1972). On present data this figure was the maximum to be reached, because in the Cretaceous the number of infested species had apparently declined and very few examples are known from the Cainozoic (Rathbun, 1916). This pattern could represent a pioneer stage (Oxfordian and Kimmeridgian) followed by a virulent spread among unresistant species (Tithonian). The post-Jurassic decline might then represent the evolution of resistant adaptations in many species. This hypothesis awaits rigorous testing, but a similar proposal was made by Brett (1978) in correlating diversification of pit-forming epizoans with adaptive radiations of their crinoid hosts, followed by the development of biochemical defences.

Several workers on these parasitic isopods have cited percentages of infested individuals in their collections; some values being 2 % (Bachmayer, 1955) and 3·82 % (Housa, 1963) for several species, and 1·33 % for a single species (Radwanski, 1972). It seems unlikely that single populations are being sampled and these percentages are open to the usual criticisms of palaeoecological data such as preferential removal or destruction of distorted carapaces, collector bias, etc. Nevertheless, these figures are valuable as they may represent averaged values for considerable periods of time that even out fluctuations in infestation percentages in the individual populations. Significantly, perhaps, the figures quoted by Bachmayer (1955) and Housa (1963) are from rocks of about the same age, apparently deposited in a similar environment. The discrepancy between these two percentages may be a genuine, albeit unexplained, ecological phenomenon (Housa, 1963).

(6) *Annelids and molluscs*

Among the most widely cited examples of fossil parasitism are a variety of cysts, galls and other disturbances in Palaeozoic and Mesozoic echinoderms which are attributed to myzostomids (Graff, 1885; Yakovlev, 1922; Moodie, 1923; Ehrenberg, 1933*a*; Roman 1952, 1953; Warn, 1974). Myzostomids are peculiar annelids, some of which produce galls and cysts in modern echinoderms. There is, however, considerable doubt as to whether many of these fossil identifications are correct (Franzen, 1974), although undoubted myzostomid activity does appear to date back to the Carboniferous (Pl. 2E) (Welch, 1976), or perhaps the Devonian (Ehrenberg, 1933*b*). Amongst other annelids the only known fossil leeches are from the Upper Jurassic of Germany, but they appear to have had a limited potential for parasitism (Kozur, 1970).

Cup and bowl-like excavations in echinoderms have been ascribed to parasitic gastropods, (Pls 1 B and 2 D, F) with examples known from both the Palaeozoic (Sieverts-Doreck, 1963; Paul, 1971; Frest, Mikulic & Paul, 1977) and Mesozoic (Mercier, 1931). These identifications remain at best tenuous, especially as the skeleton is rarely completely penetrated through to the coelom (Frest *et al.*, 1977; Brett, 1978).

More convincing examples of gastropod parasitism are known from Cretaceous echinoids (Saint-Seine, 1951; Szorenyi, 1955), including characteristic excavations in the spines of *Stereocidaris* (Tasnadi-Kubacska, 1962).

(7) *Parasites of uncertain affinities*

In a number of fossils showing parasitic attack there is either a lack of diagnostic detail or, more significantly, an absence of modern analogues that forbids assignment of a gall or tumour to a specific group of parasites. Fossil echinoderms have been one of the richest sources for this material of uncertain affinity (Pls 1 A, C and 2 C, D, F). This should not be regarded as some mysterious preference by parasites and other epizoans for echinoderms. It is simply a reflection of the reaction engendered in the stereom tissue by attachment or attack (Franzen, 1974). In crinoids such reactions were evidently most pronounced in the stems, whereas the slower rate of secretion in the calyx produced less obtrusive disturbances (Brett, 1978). As noted above, many identifications of the parasitic activities of myzostomids and gastropods seem dubious and non-committal comments are often preferable (Etheridge, 1880; Clark, 1921; Franzen, 1974; Brett, 1978). Furthermore, while the galls and excavations in echinoderms traditionally have been regarded as being of parasitic origin, this has recently been questioned (Franzen, 1974; Brett, 1978).

There are several examples that may be reasonably attributed to parasitism. These include double-pored galls (*Schizoproboscina*) which were described from the arms of Carboniferous crinoids (Yakovlev, 1939). More significant is the discovery of the enigmatic phosphatic organism *Phosphannulus* within perforated swellings of crinoid stems between Upper Ordovician and Lower Permian in age

(Pl. 1 E, F) (Welch, 1976). They were previously described as remnants of myzostomids (Warn, 1974), but the form of *Phosphannulus* as a tube with a proximally flared section shows that this is untenable (Welch, 1976). There is good evidence that *Phosphannulus* attached itself to the stem and in provoking the host reaction sank inwards, maintaining a connection with the outside world, and established contact with the central canal of the stem (Warn, 1974; Welch, 1976). *Phosphannulus* was first described from isolated specimens released by digesting limestones (Cambrian to Devonian) in acetic acid (Müller, Nogami & Lenz, 1974). While it is apparent that these specimens were usually attached to a hard substrate, it is desirable to confirm whether *Phosphannulus* could occur as a free-living form.

Malformations in trilobites may be ascribed tentatively to parasitism (Schmidt, 1906; Richter & Richter, 1934), but many are more likely to have arisen during moulting. Galls and other swellings have been described from Cambrian (Snajdr, 1978), Ordovician (Pl. 2 A, B) (Ludvigsen, 1979, personal communication) and possibly Silurian and Devonian (Snajdr, 1978) trilobites. The nature of the original parasites is uncertain. Ludvigsen (1979) proposed a myzostomid origin for one such gall (Pl. 2 A), but this is uncertain. Snajdr (1978) described several types of deformation attributable to parasites, including two types of gall from Middle Cambrian paradoxid trilobites. Interestingly, Snajdr (1978) traced an ontogenetic development of the galls in the various moult stages of the trilobite and this approach could throw more light on the affinities of the parasites. The parasites appeared to have lived beneath the exoskeleton and so survived the moulting. Neither type of gall, however, shows very specific siting. A specific location for possible parasites in trilobites is known for the meandering vermiform traces, referred to as a nematode (*Cadella flexuosa*) by Lamont (1975), found on the spines of Lower Cambrian trilobites from north-west Scotland (Peach, 1894). Further cases of supposed worm infestation in Cambrian and later trilobites were noted by Lamont (1975), but further investigations are certainly necessary. These examples of parasitism are especially significant because, being Cambrian, they occur in some of the earliest metazoan faunas (Table 1) and because amongst the arthropods trilobites show a number of primitive characteristics.

Trilobites themselves have also been regarded as parasitic, especially the Cambro-Ordovician agnostids (M'Coy, 1849; Bergström, 1973). The unusual exoskeleton of agnostids in comparison with other trilobites suggests that the soft-part anatomy was correspondingly aberrant and the mode of feeding may have been basically different (Robison, 1972). The apparent absence of suitable hosts and the prolific abundance of agnostids in many rocks show, however, that more convincing evidence is required if this hypothesis is to be supported.

A variety of structures has also been described in the extinct colonial graptolites which, to judge from the structure of their skeleton, were closely related to hemichordates such as *Rhabdopleura* (Crowther & Rickards, 1977). Blister-like structures in some pelagic graptolites of Ordovician and Silurian age may possibly be of parasitic origin (Urbanek, 1958; Jackson, 1971; P. R. Crowther, personal communication). More significant is the rare occurrence in several genera of sessile Ordovician graptolites (dendroids and tuboids) of vermiform tubes that extend from the colony, often expanding in diameter and opening distally. These tubes were described by Kozlowski (1970) as 'tubothecae' (Pl. 2, G) and were regarded as the product of parasitic polychaetes. A tubotheca can either arise from the opening of one of the zooids of the colony or directly from the graptolite wall. This suggests that on settling the parasite may not have simply attached itself to a zooid, but entered the colony (possibly during feeding) and sometimes migrated to rupture the colony wall some distance from its point of entry. Peter Crowther (City Museum and Art Gallery, Peterborough) has recently re-studied tubothecae and has very kindly allowed me to quote some of his unpublished observations. The normal graptolite skeleton was secreted as 2 layers, a fusellar layer and an outer, and sometimes inner, cortical layer. The cortex was secreted in swathes or 'bandages' of collagenous fibrils which evidently were plastered on to the existing skeleton by the pre-oral cephalic disc of the zooids when they extended from the thecal openings (Crowther & Rickards, 1977). Kozlowski (1970) recognized a thick cortical layer in the tubothecae which he regarded as encapsulating the leathery tube of the parasitic polychaete. Detailed re-study by Crowther using electron microscopy has shown that the tubotheca are indeed composed of immensely thickened cortex, although there is no evidence for the leathery tube. Crowther concludes that a parasite of uncertain affinities, whose vermiform shape is mimicked by the tubotheca, attacked the colony and may have established a connection with the central stolon that joined all the graptolite zooids. In a reaction to the parasite, adjacent zooids plastered a callus of cortex around the intruder. The persistence of the distal opening of each tubotheca suggests, however, that the parasite survived. Another example of an Ordovician association between a graptolite and vermiform animal, named *Helicosyrinx*, was documented by Kozlowski (1967). *Helicosyrinx* coiled tightly around the graptolite and was identified as a probable phoronid. Detailed ultrastructural work is required to verify that this association was not parasitic.

In addition to the examples noted above a number of other groups have traces that may represent parasitic invasion. These include borings and other disturbances in conodonts (Müller & Nogami, 1972), foraminifers (Tasnadi-Kubacska, 1962) and corals (Geczy, 1954).

OTHER FOSSIL ASSOCIATIONS AND THEIR RELEVANCE TO PARASITOLOGY

In addition to parasitism a wide variety of other associations between fossil species are known, some of which appear to verge on parasitism. The tendency for inter-specific associations to form evidently dates back to the earliest stages of metazoan history. *Cloudina* is a worm tube from the late Precambrian of Namibia and is one of the earliest known metazoans with hard parts (Germs, 1972). There are 2 species, *C. hartmannae* and *C. riemkeae*, and significantly the latter often grew epizoically on the former species.

It appears that the non-parasitic associations could throw considerable light on the history and nature of host–parasite relations. One such aspect is host-specificity. A number of examples of host-specific commensals have been described in the fossil record. For example, a strong preference for certain Devonian spiriferid brachiopods has been noticed in some boring (Thayer, 1974) and encrusting (Richards, 1974) commensals. Regnell (1966) commented on a similar selectivity of the extinct echinoderm group of edrioasteroids for host substrates, and Paul (1971) noted that amongst a diverse fauna of Silurian echinoderms only certain cystoids were attacked by boring organisms. Host specificity between commensal hydroids and serpulid worms in the Jurassic was documented by Scrutton (1975). It seems probable that many, although certainly not all, examples of infestation and attack in the fossil record are similarly specific to one or a few species. Reports of host-specificity are, however, rarely dynamic in that the origins and subsequent evolution of an association are often obscure. A notable attempt to explore the temporal aspect of host-specificity was Brett's (1978) analysis of epizoan excavations, some of which may have been parasitic, in Silurian and Devonian crinoids. In addition to noting pronounced host-specificity Brett (1978) was able to document the persistence of one epizoan type from the Upper Silurian genus *Ichthyocrinus* to its descendant Middle Devonian genus *Synaptocrinus* over a period of perhaps 30 million years. Another epizoan ranged from the Middle Devonian to Upper Carboniferous in other crinoids. These cases of co-evolution deserve much more detailed study, as there seems little reason to think that the behaviour of host-specificity in other parasites is fundamentally different.

One of the most famous commensal associations in the Palaeozoic is between several genera of coprophagous (faecal feeders) platyceratid gastropods and crinoids (Bowsher, 1955), or more rarely the extinct blastoids (Levin & Fay, 1964) and cystoids (Clarke, 1921; Bowsher, 1955). The gastropods lived over the anus of the echinoderm and there is clear evidence that the association often persisted over much of the life-span of the individuals, with the rim of the gastropod shell becoming closely moulded to irregularities in the crinoid surface. In the context of this review the platyceratid–echinoderm association presents several interesting points. A degree of host-specificity has been recognized (Clarke, 1921; Lane, 1973), although phenotypic variability sometimes makes specific identifications difficult (Bowsher, 1955). The origins of the association are uncertain, although the ancestors of the platyceratids were presumably free-living. Lane (1978) commented that in some early examples the gastropods appear not to have had a specific position on the crinoids and that an invariable location over the anus was a slightly later development. This apparent development in an association is in marked contrast to many other fossil associations where the appearance is abrupt and without apparent ancestry. Any fossils that show the progressive establishment of an association deserve close study.

The geological range of the platyceratid–echinoderm association is enormous as it ranges from Middle Ordovician to Permian, with the principal genus *Platyceras* occurring between the Middle Silurian and Upper Permian. This is a period of about 170 million years, about two and a half times the length of the Cainozoic, during which time there was no basic change in the nature of the association.

Despite changes in the host taxa this remarkably robust relationship only ended in the widescale mass extinction at the end of the Permian. Another commensal association with a similar longevity was noted by Scrutton (1975) between one or a few species of hydroids and a wider variety of serpulids, with a range from Jurassic to Pliocene. It seems possible that some parasitic associations may have also shown a similar persistence with a simple lineage of parasites infesting a more diverse succession of hosts. This accords with the parasitological 'rule' of parasite evolution being slower than host evolution (Manter, 1967).

A final point about the former association is that although platyceratid gastropods apparently had the option of becoming parasitic, with immediate access to the internal soft tissues of the echinoderm, they remained as commensals (cf. Clarke, 1921). Many parasitic associations may have started as more benign unions, but there is no evidence that associations such as commensalism are necessarily destined to shift gradually into parasitism.

PARASITES IN THE HISTORY OF LIFE

In the history of life both terrestrial and marine faunas have changed in overall character with time. In a general way the character of modern biotas may be seen to emerge in the fossil record as one approaches the Recent. In addition to this overall and gradual process the history of life has been punctuated by major upheavals and revolutions, affecting a smaller or larger proportion of the entire biota, that have been separated by longer intervals where the component faunas and floras have shown greater stability and uniformity of character.

With respect to the history of parasitism these observations bear on two points. Firstly, the foundations of modern biotas were largely laid in the late Mesozoic and early Cainozoic, so it may be permissible to extrapolate knowledge of modern parasites back to these times. Baxter's (1975) remark that the distinctiveness of ancient floras were probably matched by their associated fungi, a group that is not considered here, can be extended to a general principle that the alien appearance, to our eyes, of the older faunas was probably reflected in the parasitic component. The preponderance of extinct groups in earlier times, especially in the Palaeozoic, suggests that some of their parasites may have been as distantly related to modern forms as their hosts. It may be unlikely that any of these parasites belonged to extinct phyla, although a number of Palaeozoic animals appear to fall into this category. Many of these parasites, however, probably represent now extinct classes.

An exception to these generalizations is the celebrated 'living fossils', which have persisted to the present day with little morphological change for many millions of years. Well known examples amongst the vertebrates are the chimaeroid fish, coelocanths, lung-fish and the tuatara 'lizard'. Similar examples are known amongst the invertebrates. A notable feature of 'living fossils' is that while most show an early evolutionary expansion, the rest of their history is marked by extremely low diversities (Stanley, 1979) which presumably indicates a fairly linear and simple phylogeny. Unfortunately, a general review of the parasites of 'living fossils' does not appear to have been compiled. Such a synthesis would be welcome. It is clear, however, that at least some 'living fossils' do possess unusual parasites, for example, the gyrocotylideans in chimaeroid fish. These parasites may be relicts of once larger groups that infested the hosts during their hey-day, and although giving rise to other major groups of parasitic platyhelminths (Llewellyn, 1965) have themselves shown little change. Their evolution may run parallel to the simple lineage of the host. Other 'living fossils' appear not to carry this burden of archaic parasites and seem to have been open to infections by successive waves of more recent parasites. As only some 'living fossils' appear to retain comparably ancient parasites it will be interesting to see how this discrepancy arose in groups which, from an evolutionary perspective, are similar.

The second point is that while the crises and revolutions in the history of life presumably had their proportional effect on parasites, it is difficult as yet to identify specific effects. Sepkoski (1979) has delimited the unique character of the Cambrian marine faunas in comparison with the later Palaeozoic faunas which arose from a major adaptive radiation during the Ordovician. I suspect that this uniqueness was characteristic also of Cambrian parasites, although the scanty evidence (for example, Snajdr, 1978) is inconclusive. In the mid-Mesozoic there was another repatterning of marine life with the emergence of a more modern fauna (Vermeij, 1977). It seems plausible that the appearance of some marine parasites such as the ascothoracican barnacles and epicaridean isopods is part of this phenomenon. The mass extinctions that punctuate the fossil record (Table 1) presumably took their toll of parasites, but no direct evidence is available. The termination of the platyceratid gastropod–crinoid association at the end of the Permian, having weathered many more minor extinctions, is a hint of the major remodelling of associations that must have occurred both during the extinctions and the ensuing adaptive radiations, as surviving groups re-occupied vacant ecospace.

A still broader problem is whether host–parasite relationships have changed fundamentally with geological time. Moodie (1923) indicated that early Palaeozoic faunas were free of disease, with parasitism appearing only in the Devonian; a view which was based on limited data. The occurrence of host-specific associations dating back at least to the Silurian and evidence for parasites from the Cambrian argues for parasitism being established early in metazoan history. It is interesting to speculate, however, whether the low level of interaction that seems to be a feature of the relatively simple ecosystems of Precambrian and Cambrian metazoan faunas was reflected in a restriction of parasitism.

In the absence of a coherent fossil record the evidence for the evolution of parasites must come from neontological studies. A temporal perspective, however, can only come in many cases from analysing the better known fossil record of the hosts. The connection between the evolution of hosts and parasites is fraught with problems, as is apparent from the protracted discussions surrounding the various parasitological 'rules'. Any correlation between host and parasite in the geological record must assume that the existing degree of host-specificity is a rough reflection of its prior history, with the greater degree of specificity indicating a higher fidelity in reconstructing phylogenies. This attitude has been adopted by many parasitologists (for example, Manter, 1967), although it may not always be justified (Ewers, 1963; Inglis, 1971).

In the case of single host parasites, such as the monogenean platyhelminths, the geological range (if known) of the one host may give the only clue on the age of association. The pronounced host-specificity of monogeneans could make such an approach viable. A greater degree of resolution may come in parasites with 2 or more hosts, by comparing the geological ranges of the host groups to discover whether these temporal ranges show a discordant or concordant relationship. The fossil record of some hosts, often of the most interest to parasitologists, is too poor for useful analysis; but for others, for example, vertebrates and molluscs, the record is tolerably good. These latter two groups act as the intermediate and final hosts of digenean trematodes. The distribution of trematodes in various mollusc families, mostly of freshwater and terrestrial gastropods, is reviewed by Ewers (1964). The geological ranges of some of these families is summarized by Solem (1979). The gastropod family Enidae, for instance, ranges from the Carboniferous and presently hosts two families of trematode. It is conceivable that parasitism of the mollusc first occurred in the Palaeozoic. Infestation of the mammal and bird hosts probably only dates from the Upper Jurassic – Lower Cretaceous, so that the discordance between the geological ranges of the two hosts could be as much as 200 million years. Granted that parasitism by the trematodes originally arose in association with Palaeozoic molluscs it might be argued that prior to the mid-Mesozoic the molluscs were the sole hosts (Llewellyn, 1965; Manter, 1967). This could explain the much greater degree of host-specificity. Alternatively, another vertebrate group may have acted as an earlier final host. In the majority of cases a concordant relationship between mollusc and vertebrate hosts may be suspected from their similar geological ranges. In examples of an apparent discordant relationship further illumination on the host–parasite relationship may arise if a potential host with a suitable ecological and geographic, although not necessarily phylogenetic, pedigree is found to precede the host with the smaller age range. A few examples of an apparent ecological succession between unrelated groups are known from the fossil record, although it is not known if transfer of host-specific symbionts such as epizoic commensals ever took place.

PARASITES AND MACRO-EVOLUTION

(1) *Speciation*

Problems of speciation in parasites have received attention (Inglis, 1965, 1971), but many aspects of macro-evolution in parasites have been more neglected. The recent synthesis by Stanley (1979) makes a brief examination of macro-evolution relevant here, especially as renewed interest in macro-evolutionary models has received considerable impetus from an analysis of the fossil record. There is only space to consider a few facets of macro-evolution and readers are recommended to the excellent text of Stanley (1979) for further insights.

Stanley (1979) regards the species as the basic unit in macro-evolution, and furthermore believes that micro- and macro-evolutionary processes are largely decoupled and do not form a continuum. This decoupling means that much of the evolutionary research by biologists describes the 'fine tuning' of a species-adaptive capability and sheds little light on the origin of species and the evolution of major taxonomic categories, subjects upon which the fossil record can supply information. Speciation has traditionally been regarded as the gradual and cumulative evolution via intermediate forms that provide a continuum between ancestral and descendant species. Differentiation between such chronospecies, therefore, would be largely an arbitrary choice of sub-dividing a continuous lineage. It has long been realized that there is a conspicuous absence of such gradualistic lineages in the fossil record with most fossil species appearing suddenly in the rock record. This anomaly between what is observed and the model of gradualistic evolution was interpreted by the stratigraphic column being riddled with minor gaps that removed the intermediate forms, so explaining the abrupt appearance of most fossil species. This view has been challenged by a new model, that of punctuated equilibria, developed by N. Eldredge and S. J. Gould. This model is an extension into the fossil record of the earlier evolutionary hypothesis which saw many species arising rapidly from very small, possible peripheral, populations. One testable consequence of the punctuated equilibria model is that once established, the character of a species is maintained with very little change for periods of time far in excess of the time taken for speciation. In many fossil species this conservatism is shown by the negligible change in morphology over enormous periods of time. This equilibrial aspect of the model is regarded as reflecting the fact that once a species is established it is a very stable entity adapted to its environment and intolerant to any marked shift in the genotype. Small populations, perhaps only a few individuals, may 'experiment' with major shifts which are usually unsuccessful, but in rare instances may lead to speciation. Significantly there are a number of factors that lead to the establishment of small populations in parasites with little gene flow between them (Price, 1977).

The model of punctuated equilibria explains many aspects of macro-evolution and has a direct bearing on the origin and evolution of parasites. There seems little reason to suppose that parasites speciate in a more gradualistic fashion (cf. Inglis, 1965, 1971) than other species, although at first sight coevolution would seem to require a gradual adjustment between host and parasite. This is because of the implication of close tracking between their evolutionary histories which would seem to be precluded by punctuated breaks disrupting the pattern. Stanley (1979), however, has emphasized the opportunistic nature of speciation with small mutant populations continuously 'testing' potentially new niches. Parasites appear to have high speciation rates (Price, 1977). It seems plausible, therefore, that if the existing parasites are maladapted during host speciation one or more parasite populations would probably speciate and continue to infect the new species of host. Whether this process continued through the complex branching of many clades, usually simplified in the imagination to a few simple lineages, is uncertain.

(2) *Adaptive radiations*

Adaptive radiations, difficult to explain in a gradualistic context, readily fall into the punctuated-equilibria model (Stanley, 1979). Many adaptive radiations either arise from an evolutionary breakthrough or from the vacation of ecospace following the extinction of one group and its replacement by another. In the former case an adaptive radiation of a parasitic group (Price, 1977) could arise at any time and would be largely independent of its hosts. The radiation might stem from the evolution of a resistant larva or a physiological mechanism to combat host defences. In relation to the latter case there does not seem to be any clear evidence for the adaptive radiation of one group of parasites arising from the demise of another, independent of the evolutionary fortunes of the host. A renewed study of fossil associations may reveal such an event. It seems very likely, however, that some adaptive radiations in parasites stem directly from those of their hosts. Brett (1978) has commented on an apparent correlation in the Palaeozoic of epizoan abundance, perhaps including ectoparasites, with adaptive radiations in the host crinoids. A significant feature of adaptive radiations is that the basic characters of a group usually emerge at a very early stage. Subsequent evolution tends to express variations on a theme, rather than any radically new forms. It seems probable that the major steps involved in the evolution of a new parasitic group probably appeared over a short geological time rather than arising in a piecemeal fashion. The rapidity and extent of evolution during an adaptive radiation shows that there may be no need to postulate immense periods of time to account for specialization in parasites. It seems questionable to what extent one can assume that specialized parasites show a greater antiquity than groups apparently less adapted to parasitism. Factors such as the pre-adaptive potential and plasticity of development may account more readily for differences in the degree of parasitism than relative timing in their evolutionary history.

I offer particular thanks to A. J. Boucot and W. D. I. Rolfe for freely sharing information and bringing my attention to a wide range of literature. J. N. Thomas and W. D. I. Rolfe also critically read the manuscript. I thank W. G. Chaloner, C. Franzen, K. Larsson, R. Ludvigsen, P. Crowther, P. S. Gooch, C. R. C. Paul, R. J. G. Savage, P. A. Selden, A. W. A. Rushton and S. F. Morris for helpful information. C. Franzen, J. R. Welch, R. Ludvigsen P. Crowther, and C. E. Brett, kindly made available photographs and gave their permission to use published material. Acknowledgement is made to the Paleontological Society, Palaeontological Association, the Geological Survey of Canada and the editors of *Lethaia* for use of material published by them. My wife Zoë patiently translated a number of German papers.

REFERENCES

ALVIN, K. L., BARNARD, P. D. W., HARRIS, T. M., HUGHES, N. F., WAGNER, R. H. & WESLEY A. (1967). Gymnospermophyta. In *The Fossil Record* (ed. W. B. Harland), pp. 247–268. London: Geological Society of London.

BACHMAYER, F. (1948). Pathogene Wucherungen bei jurassischen Dekapoden. *Sitzungsberichte der österreichischen Akademie der Wissenschaften*, Abt. 1 **157**, 263–6.

BACHMAYER, F. (1965). Die fossilen Asseln aus den Oberjuraschichten von Ernstbruun in Niederösterreich und von Stramberg in Mahren. *Sitzungsberichte der österreichischen Akademie der Wissenschaften*, Abt. 1 **164**, 255–73.

BAXTER, R. W. (1975). Fossil fungi from American Pennsylvanian coal balls. *Paleontological Contributions, University of Kansas* **77**, 1–6.

BERGSTRÖM, J. (1973). Organization, life, and systematics of trilobites. *Fossils and Strata* **2**, 1–69.

BOWSHER, A. L. (1955). Origin and adaptation of platyceratid gastropods. *Paleontological Contributions. University of Kansas* **5**, 1–11.

BRETT, C. E. (1978). Host-specific pit-forming epizoans on Silurian crinoids. *Lethaia* **11**, 217–32.

BROWN, R. W. (1940). Fossil pearls from the Colorado group of Western Kansas. *Journal of the Washington Academy of Sciences* **30**, 365–74.

CHATTERTON, B. D. E. (1975). A commensal relationship between a small filter feeding organism and Australian Devonian spiriferid brachiopods. *Paleobiology* **1**, 371–8.

CLARKE, J. M. (1921). Organic dependence and disease. *Bulletin of the New York State Museum* **221, 222**, 1–113.

CONWAY MORRIS, S. (1977). Fossil priapulid worms. *Special Papers in Palaeontology* **20**, 1–95.

CRESSEY, R. & PATTERSON, C. (1973). Fossil parasitic copepods from a Lower Cretaceous fish. *Science* **180**, 1283–5.

CROMPTON, D. W. T. (1975). Relationships between Acanthocephala and their hosts. *Symposia of the Society for Experimental Biology* **29**, 467–504.

CROWTHER, P. & RICKARDS, B. (1977). Cortical bandages and the graptolite zooid. *Geologica et Palaeontologica* **11**, 9–46.

DEAN, B. (1902). The preservation of muscle-fibres in sharks of the Cleveland Shale. *American Geologist* **30**, 273–8.

DOLLFUS, R. PH. (1950). Liste des némathelminthes connus à l'état fossile. *Bulletin de la Société de France. Compte rendu Sommaire des Seances*, **20**, Ser. 5, 82–5.

DUBININ, V. B. (1948). Pleistocene fleas (Anoploura) and nematodes found in a study of fossil sousliks from the basin of the River Indiquirka. (North-east Siberia). *Doklady Akademii nauk SSSR* **62**, 417–20. (In Russian).

DUBININA, M. N. (1972). The nematode *Alfortia edentatus* (Looss, 1900) from the intestine of an Upper Pleistocene horse. *Parazitologia* **6**, 441–3. (In Russian).

EHRENBERG, K. (1933*a*). Über eine bemerkenswerte Crinoidenstiel-Deformität. *Palaeobiologica* **5**, 201–10.

EHRENBERG, K. (1933*b*). Ein mutmasslicher Fall von Parasitismus bei der devonischen Crinoidengattung *Edriocrinus*. *Biologia generalis* **9**, 85–96.

ETHERIDGE, R. (1880). Observations on the swollen condition of Carboniferous crinoid stems. *Proceedings of the Natural History Society of Glasgow* **9**, 19–36.

EWERS, W. H. (1964). An analysis of the molluscan hosts of the trematodes of birds and mammals and some speculations on host-specificity. *Parasitology* **54**, 571–8.

FERRIS, V. R., GOSECO, C. G. & FERRIS, J. M. (1976). Biogeography of free-living soil nematodes from the perspective of plate tectonics. *Science* **193**, 508–10.

FÖRSTER, R. (1969). Epökie, Entökie, Parasitismus und Regeneration bei fossilen Dekapoden. *Mitteilungen der Bayerischen Staatissmus palaontologischen, historischen und geologischen* **9**, 45–59.

FRANZEN, C. (1974). Epizoans on Silurian-Devonian crinoids. *Lethaia* **7**, 287–301.

FREST, T. J., MIKULIC, D. G. & PAUL, C. R. C. (1977). New information on the *Holocystites* fauna (Diploporita) of the Middle Silurian of Wisconsin, Illinois, and Indiana. *Fieldiana: Geology* **35**, 83–108.

GECZY, B. (1954). Cyclolites (Anth.) tanulmanyok. *Geologica Hungarica, Seria palaeontologica* **24**, 1–180.

GERMS, G. J. B. (1972). New shelly fossils from Nama Group, South West Africa. *American Journal of Science* **272**, 752–61.

GLENISTER, B. F., KLAPPER, G. & CHAUFF, K. M. (1976). Conodont pearls? *Science* **193**, 571–3.

GOLVAN, Y. J. (1958). Le phylum des Acanthocephala. Première note. Sa place dans l'echelle zoologique. *Annales de parasitologie humaine et comparée* **33**, 538–602.

GOOCH, P. S. (1975). Helminths in archaeological and pre-historic deposits. *Annotated Bibliography of the Commonwealth Institute of Helminthology*, **9**, 1–15.

GRAEBER, K. & STORCH, V. (1978). Elektronenmikroskopische und morphometrische Untersuchungen am Integument der Acanthocephala (Aschelminthes). *Zeitschrift fur Parasitenkunde* **57**, 121–35.

GRAFF, L. VON (1885). Ueber einige Deformitaten an fossilen Crinoiden. *Palaeontographica* **31**, 183–91.

HEYDEN, C. (1860). *Mermis antiqua*, ein fossiler Eingeweidewurm. *Stettiner entomologische Zeitung* **21**, 38.

HEYDEN, C. (1862). Gliederthiere aus der Braunkohle des Niederrhein's, der Wetterau und der Rohn. *Palaeontographica* **10**, 62–82.

HOUSA, V. (1963). Parasites of Tithonian decapod crustaceans (Stramberk, Moravia). *Sbornik Ustredniho ustavu geologickeho* **28**, 101–14.

INGLIS, W. G. (1965). Patterns of evolution in parasitic nematodes. In *Evolution of Parasites, 3rd Symposium of the British Society for Parasitology* (ed. A. E. R. Taylor), pp. 79–124. Oxford: Blackwell.

INGLIS, W. G. (1971). Speciation in parasitic nematodes. In *Advances in Parasitology*, vol. 9 (ed. B. Dawes) pp. 185–223. London and New York: Academic Press.

JACKSON, D. E. (1971). Development of *Glyptograptus hudsoni* sp. nov. from Southampton Island, North-West Territories, Canada. *Palaeontology* **14**, 478–86.

KEVAN, P. G., CHALONER, W. G. & SAVILE, D. B. O. (1975). Interrelationships of early terrestrial arthropods and plants. *Palaeontology* **18**, 391–417.

KOZLOWSKI, R. (1967). Sur certains fossiles Ordoviciens à test organique. *Acta palaeontologica polonica* **12**, 99–132.

KOZLOWSKI, R. (1970). Tubotheca – a peculiar morphological element in some graptolites. *Acta palaeontologica polonica* **15**, 393–410.

KOZUR, H. (1970). Fossile Hirudinea aus dem Oberjura von Bayern. *Lethaia* **3**, 225–32.

LAMONT, A. (1975). Cambrian trilobites from the Pass of Leny, Perthshire, Scotland. *Scottish Journal of Science* **1**, 199–215.

LANE, N. G. (1973). Paleontology and paleoecology of the Crawfordsville fossil site (Upper Osagian, Indiana). *University of California Publications in Geological Sciences* **99**, 1–141.

LANE, N. G. (1978). Mutualistic relations of fossil crinoids. In *Treatise on Invertebrate Paleontology*, (ed. R. C. Moore and C. Teichert), pp. T 345–347. Lawrence: University of Kansas Press and Geological Society of America.

LARSSON, K. (1979). Silurian tentaculitids from Gotland and Scania. *Fossils and Strata* **11**, 1–180.

LEVIN, H. L. & FAY, R. O. (1964). The relationship between *Diploblastus kirkwoodensis* and *Platyceras* (*Platyceras*). *Oklahoma Geology Notes* **24**, 22–9.

LESQUEREUX, L. (1892). The flora of the Dakota Group. *Monograph of the United States Geological Survey* **17**, 1–400.

LLEWELLYN, J. (1965). The evolution of parasitic platyhelminths. In *Evolution of Parasites, 3rd Symposium of the British Society for Parasitology*, (ed. A. E. R. Taylor) pp. 47–78. Oxford: Blackwell.

LUDVIGSEN, R. (1979). A trilobite zonation of Middle Ordovician rocks, southwestern District of Mackenzie. *Bulletin, Geological Survey of Canada* **312**, 1–98.

MACKINNON, D. I. & BIERNAT, G. (1970). The probable affinities of the trace fossil *Diorygma atrypophilia*. *Lethaia* **3**, 163–72.

MADSEN, N. & WOLFF, T. (1965). Evidence of the occurrence of Ascothoracica (parasitic cirripedes) in Upper Cretaceous. *Meddelelser fra Dansk geologisk Forening* **15**, 556–8.

MANTER, H. W. (1967). Some aspects of the geographical distribution of parasites. *Journal of Parasitology* **53**, 1–9.

MARGARA, J. (1946). Existence de zoothylacies chez des Clypéastres (Echinodermes) de l'Helvétien du Proche-Orient. *Bulletin du Muséum national d'histoire naturelle* **18**, Ser. 2 423–7.

M'COY, F. (1849). On the classification of some British fossil Crustacea, with notices of new forms in the University Collection at Cambridge. *Annals and Magazine of Natural History* **4**, Ser. 2, 161–79 and 392–414.

MENGE, A. (1866). Ueber ein Rhipidopteron und einige andere im Bernstein eingeschlossene Thiere. *Schriften der Naturforschenden Gesellschaft in Danzig* **3**, Ser. 8, 12–15.

MERCIER, J. (1931). Note sur des crinoides parasites de la couche à *Leptaena* (Toarcien) de May-sur-Orne. *Bulletin de la Société linnéenne de Normandie* **3**, 12–15.

MERCIER, J. (1936). Zoothylacies d'échinide fossile provoquées par un crustacé: *Castexia douvillei* nov.gen., nov.sp. *Bulletin de la Société geologique de France* **6**, Ser. 5, 149–54.

MEYER, A. (1933). Acanthocephala. In H. G. Bronns *Klassen und Ordnungen des Tierreichs*, Band IV, 2, II, pp. 582.

MOODIE, R. L. (1923). *Paleopathology. An introduction to the Study of Ancient Evidences of Disease*. Urbana: University of Illinois Press.

MOUSSA, M. T. (1969). Nematode fossil tracks of Eocene age from Utah. *Nematologica* **15**, 376–80.

MOUSSA, M. T. (1970). Nematode fossil trails from the Green River Formation (Eocene) in the Uinta Basin, Utah. *Journal of Paleontology* **44**, 304–7.

MÜLLER, K. J. (1979). Phosphatocopine ostracodes with preserved appendages from the Upper Cambrian of Sweden. *Lethaia* **12**, 1–27.

MÜLLER, K. J. & NOGAMI, Y. (1972). Entöken und Bohrspuren bei den Conodontophorida. *Palaontologische Zeitschrift* **46**, 68–86.

MÜLLER, K. J., NOGAMI, Y. & LENZ, H. (1974). Phosphatische Ringe als Mikrofossilien im Altpalaozoikum. *Palaeontographica*, Abt. A, **146**, 79–99.

NICHOLAS, W. L. (1971). The evolutionary origins of the Acanthocephala. *Journal of Parasitology* **57**, 84–7.

OSCHE, G. (1963). Morphological, biological, and ecological considerations in the phylogeny of parasitic nematodes. In *The Lower Metazoa. Comparative Biology and Phylogeny*, (ed. E. C. Dougherty), pp. 283–302. Berkeley: University of California Press.

PALIJ, V. M., POSTI, E. & FEDONKIN, M. A. (1979). Soft-bodied Metazoa and trace fossils of Vendian and Lower Cambrian. In *Upper Precambrian and Cambrian Paleontology of East-European Platform* (ed. B. M. Keller and A. Yu. Rozanov), pp. 49–82. Moscow: Academy of Sciences. (In Russian.)

PAN, C-T. (1963). Generalized and focal tissue responses in the snail, *Australorbis glabratus*, infected with *Schistosoma mansoni*. *Annals of the New York Academy of Sciences* **113**, 475–85.

PAUL, C. R. C. (1971). Revision of the *Holocystites* fauna (Diploporita) of North America. *Fieldiana: Geology* **24**, 1–166.

PEACH, B. N. (1894). Additions to the fauna of the Olenellus-Zone of the North-West Highlands. *Quarterly Journal of the Geological Society of London* **50**, 661–76.

PIERCE, W. D. (1960). Silicified Turbellaria from Calico Mountains nodules. *Bulletin of the South California Academy of Sciences* **59**, 138–43.

PIERCE, W. D. (1964). Three new types of invertebrates extracted from Miocene petroliferous nodules. *Bulletin of the South California Academy of Sciences* **63**, 81–5.

POINAR, G. O. (1977). Fossil nematodes from Mexican amber. *Nematologica* **23**, 232–8.

PONOMARENKO, A. G. (1976). A new insect from the Cretaceous of Transbaikalia, a possible parasite of pterosaurians. *Palaeontology Journal* **10**, 339–43.

POTONIE, H. (1893). Die Flora des rothliegenden von Thüringen. *Abhandlungen des preussischen geologischen Landesanstalt* **2**, 1–298.

PRICE, P. W. (1977). General concepts on the evolutionary biology of parasites. *Evolution* **31**, 405–20.

RADWANSKI, A. (1972). Isopod-infected prosoponids from Upper Jurassic, Poland. *Acta geologica polonica* **22**, 499–506.

RATHBUN, M. J. (1916). Description of a new genus and species of fossil crab from Port Townsend, Washington. *American Journal of Science* **41**, 344–6.

REGNELL, G. (1966). Edrioasteroids. In *Treatise on Invertebrate Paleontology*, (ed. R. C. Moore), pp. U136–173. Lawrence: Geological Society of America and University of Kansas Press.

REPETSKI, J. E. (1977). A fish from the Upper Cambrian of North America. *Science* **200**, 529–31.

RICHARDS, R. P. (1974). Ecology of the Cornulitidae. *Journal of Paleontology* **48**, 514–23.

RICHTER, R. & RICHTER, E. (1934). Missbildungen bei Scutellidae und konstruktive Konvergenzen. *Senckenbergiana* **16**, 155–60.

RIEK, E. F. (1970). Lower Cretaceous fleas. *Nature, London* **227**, 746–7.

ROBISON, R. A. (1972). Hypostoma of agnostid trilobites. *Lethaia* **5**, 239–48.

ROMAN, J. (1952). Quelques anomalies chez *Clypeaster melitensis* Michelin. *Bulletin de la Société géologique de France* **2**, Ser. 6, 3–11.

ROMAN, J. (1953). Galles de myzostomides chez les clypéastres de Turquie. *Bulletin du Muséum national d'histoire naturelle* **25**, 650–4.

ROTHSCHILD, M. (1936). Gigantism and variation in *Peringia ulvae* Pennant 1777, caused by infection with larval trematodes. *Journal of the Marine Biological Association of the United Kingdom* **20**, Ser. 5, 309–15.

SAINT-SEINE, R. (1951). Lésions et régénération chez le *Micraster*. *Bulletin de la Société geologique de France* **20**, 309–15.

SCHMIDT, F. (1906). Revision der ostbaltischen silurischen Trilobiten, nebst geognostischen Übersicht. *Mémoires de l'Academie impériale des Sciences de St-Petersbough* **10**, 1–62.

SCHRAM, F. R. (1973). Pseudocoelomates and a nemertine from the Illinois Pennsylvanian. *Journal of Paleontology* **47**, 985–9.

SCIACCHITANO, I. (1955). Su un Gordio fossile. *Monitore zoologico italiano* **63**, 57–61.

SCOTT, A. C. (1977). Coprolites containing plant material from the Carboniferous of Britain. *Palaeontology* **20**, 59–68.

SCRUTTON, C. T. (1975). Hydroid-serpulid symbiosis in the Mesozoic and Tertiary. *Palaeontology* **18**, 255–74.

SEPKOSKI, J. J. (1979). A kinetic model of Phanerozoic taxonomic diversity II. Early Phanerozoic families and multiple equilibria. *Paleobiology* **5**, 222–51.

SIEVERTS-DORECK, H. (1963). Über Missbildungen bei *Cupressocrinus elongatus* aus dem Mitteldevon der Eifel. *Decheniana* **115**, 239–44.

SLITER, W. V. (1971). Predation on benthic foraminifers. *Journal of Foraminiferal Research* **1**, 20–9.

SNAJDR, M. (1978). Anomalous carapaces of Bohemian paradoxid trilobites. *Sbornik Ustredniho ustavu geologickeho* **20**, 7–31.

SOLOV'YEV, A. N. (1961). A parasite in late Jurassic sea-urchins. *Paleontologicheskii zhurnal*, **4**, 115–19. (In Russian.)

SOLEM, A. (1979). Biogeographic significance of land snails, Paleozoic to Recent. In *Historical Biogeography, Plate Tectonics, and the Changing Environment* (ed. J. Gray and A. J. Boucot), pp. 277–87. Corvallis: Oregon State University Press.

STANLEY, S. M. (1979). *Macroevolution, pattern and process*. San Francisco: W. H. Freeman.

STØRMER, L. (1963). *Gigantoscorpio willsi* a new scorpion from the Lower Carboniferous of Scotland and its associated preying microorganisms. *Skrifter utgitt av det Norske videnskaps-akademi i Oslo* **8**, 1–171.

STURROCK, R. F. & STURROCK, B. M. (1971). Shell abnormalities in *Biomphalaria glabrata* infected with *Schistosoma mansoni* and their significance in field transmission studies. *Journal of Helminthology* **45**, 201–10.

SUDHAUS, W. (1974). Nematoden (insbesondere Rhabditiden) des Strandanwurfs und ihre Beziehungen zu Krebsen. *Faunistisch–Okologische Mitteilungen*, **4**, 365–400.

SZORENYI, E. (1955). Echinides Crétacés de la montagne Bakony. *Geologica hungarica, seria palaeontologica* **26**, 149–281.

TASNADI-KUBACSKA, A. (1962). *Paläpathologie, pathologie der Vorzeitlichen Tiere*. Jena: Gustav Fischer Verlag.

TAYLOR, A. L. (1935). A review of the fossil nematodes. *Proceedings of the Helminthological Society of Washington* **2** (11), 47–9.

TETRY, A. (1936). Déformation parasitaire sur la tige d'un *Pentacrinus* du Lias. *Archives de zoologie expérimentale et générale* **78**, 8–12.

THAYER, C. W. (1974). Substrate specificity of Devonian epizoa. *Journal of Paleontology* **48**, 881–94.

URBANEK, A. (1958). Monograptidae from erratic boulders of Poland. *Palaeontologia polonica* **9**, 1–105.

VAN CLEAVE, H. J. (1947). The Eoacanthocephala of North America, including the description of *Eocollis arcanus*, new genus and new species, superficially resembling the genus *Pomphorhynchus*. *Journal of Parasitology* **33**, 285–96.

VERESHCHAGIN, N. K. (1975). The mammoth from the Shandrin River. *Vestnik Zoologii Kiev* **2**, 81–4. (In Russian.)

VERMEIJ, G. J. (1977). The Mesozoic marine revolution: evidence from snails, predators and grazers. *Paleobiology* **3**, 245–58.

VOIGT, E. (1938). Ein fossiler Saitenwurm (*Gordius tenuifibrosus*, n.sp.) aus der eozanen Braunkohle des Geiseltales. *Nova acta Leopoldina* **5**, 351–60.

VOIGT, E. (1957). Ein parasitischer Nematode in fossiler Coleopteren-Muskulatur aus der eozanen Braunkohle des Geiseltales bei Halle (Saale). *Paläontologische Zeitschrift* **31**, 35–9.

VOIGT, E. (1959). *Endosacculus moltkiae*. n.g.n.sp., ein vermutlicher fossiler Ascothoracide (Entomostr.) als Cystenbildner bei der Oktokoralle *Moltkia minuta*. *Paläontologische Zeitschrift* **33**, 211–23.

VOIGT, E. (1967). Ein vermutlicher Ascothoracide (*Endosacculus* (?) *najdini* n.sp.) als Bewohner einer kretazischen *Isis* aus der UdSSR. *Paleontologische Zeitschrift* **41**, 86–90.

WARN, J. M. (1974). Presumed myzostomid infestation of an Ordovician crinoid. *Journal of Paleontology* **48**, 506–13.

WELCH, J. R. (1976). *Phosphannulus* on Paleozoic crinoid stems. *Journal of Paleontology* **50**, 218–25.

WHITFIELD, P. J. (1971*a*). The locomotion of the acanthor of *Moniliformis dubius* (Archiacanthocephala). *Parasitology* **62**, 35–47.

WHITFIELD, P. J. (1971*b*). Phylogenetic affinities of Acanthocephala: an assessment of ultrastructural evidence. *Parasitology* **63**, 49–58.

YAKOVLEV, N. N. (1922). Über den Parasitismus der Wurmer Myzostomidae auf den palaozoischen Crinoiden. *Zoologisches Anzeiger* **54**, 287–91.

YAKOVLEV, N. N. (1939). Sur la découverte d'un parasite original des crinoides marins du Carbonifère. *Doklady Akademii Nauk SSSR* **22**, 146–8.

EXPLANATION OF PLATES

PLATE 1

A. Barrel-shaped swelling with a central hole in a crinoid stem. Lower Palaeozoic, locality uncertain. Original in Franzen (1974), fig. 10A.

B. Circular borings in the tegmen partition plates of the crinoid *Eucalyptocrinites caelatus*. Silurian (Rochester Shale), Middleport, New York State. Original in Brett (1978), fig. 7A.

C. Gall-like swelling (visible lower right-hand corner) associated with shallow pits on the calyx of the crinoid *E. caelatus*. Silurian (Rochester Shale), Thorold, Ontario. Original in Brett (1978), fig. 8A.

D. The priapulid *Ancalagon minor*, a possible ancestor to the acanthocephala. Cambrian (Burgess Shale), Field, British Columbia. Original in Conway Morris (1977), pl. 25, fig. 1.

E. Transverse section of the phosphatic organism *Phosphannulus* embedded in a crinoid stem. Note the proximal end of the funnel rests on a plug of stereom and the axial canal of the stem is visible beneath this plug. Permian (Foraker Limestone), Wabaunsee County, Kansas. Original of Welch (1976), pl. 1, fig. 3.

F. Transverse section of the proximal end of a *Phosphannulus* funnel embedded in a crinoid stem. Carboniferous (Haney Limestone), Crawford County, Indiana. Original of Welch (1976), pl. 1, fig. 5.

PLATE 2

A. Gall-like swelling on the head (right-hand side) of the trilobite *Ceraurus hirsuitus*. Ordovician (Esbataottine Formation), Sunblood Range, District of Mackenzie. Original of Ludvigsen (1979). pl. 14, fig. 16.

B. Gall-like swelling on the tail (left-hand side) of the trilobite *Ceraurinella nahanniensis*. Ordovician (Esbataottine Formation), North-West Territories.

C. Gall-like swelling in crinoid stem. Devonian, Tala n' Taleb, Morocco. Original in Franzen (1974), fig. 8*b*).

D. A swelling and associated pits in a crinoid stem. Silurian (Klinteberg Beds), Gotland, Sweden. Original in Franzen (1974), fig. 7A.

E. Myzostomid gall on crinoid arm. Carboniferous (Seminole Formation), Tulsa County, Oklahoma. Original in Welch (1976), text-fig. 2.

F. Pits and associated swelling in the calyx and proximal stem of the crinoid *Icthyocrinus laevis*. Silurian (Rochester Shale), Niagara Gorge, New York State. Original of Brett (1978), fig. 4A.

G. Tubotheca from a graptolite (*Dictyonema*). Ordovician, Sweden.

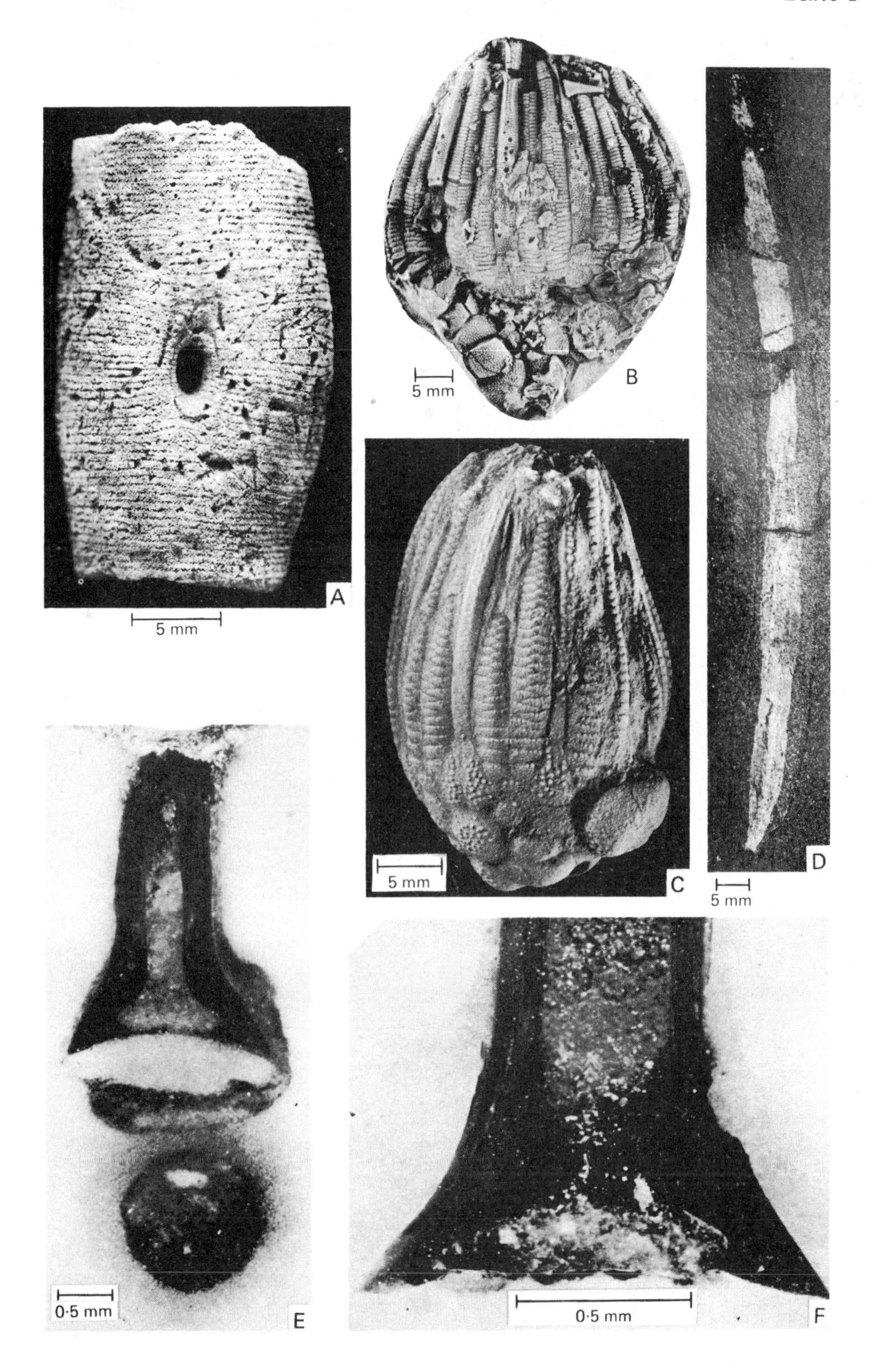
A
5 mm
B
5 mm
C
5 mm
D
5 mm
E
0·5 mm
F
0·5 mm

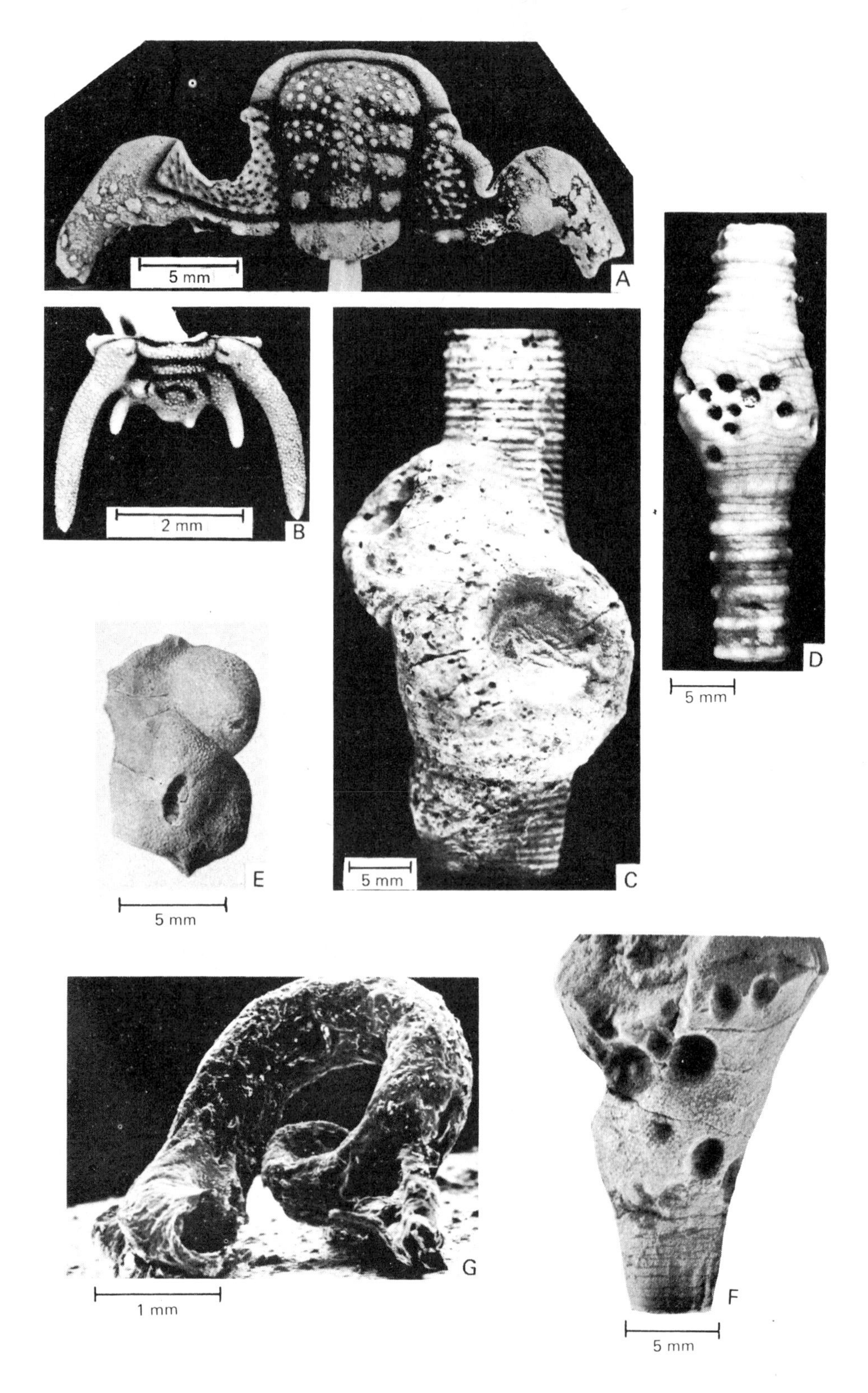
5 mm
A
2 mm
B
5 mm
C
D
5 mm
E
5 mm
G
1 mm
F
5 mm

Progress in immunization against parasitic helminths

SHEELAGH LLOYD

Department of Clinical Veterinary Medicine, University of Cambridge, Madingley Road, Cambridge

(*Accepted* 14 *November* 1980)

INTRODUCTION

Helminths are extremely numerous and cosmopolitan parasites of man and domesticated animals and they are responsible for considerable clinical and sub-clinical disease and tremendous economic loss. Thus, immunization against the parasitic helminths of man and animals would be of major health importance and to considerable economic advantage.

It is well known that animals develop some, and often a marked degree of acquired resistance to reinfection with helminths and it is on such acquired protective immunity that the development of vaccines is based. However, the immunological control of helminth infections has not yet been widely developed despite the fact that the ultimate control of many helminth diseases may be dependent on the development of effective immunoprophylaxis and despite the fact that effective vaccines are available for the control of many bacterial and viral infections.

Immunization against parasitic helminths has been considered in several reviews, recent examples being those of Clegg & Smith (1978) and Taylor & Muller (1980). Therefore, the present discussion will not encompass details as to the procedures used in attempts to immunize against parasitic infections, but will examine recent developments in the various approaches to the development of vaccines against helminth infections. In general, these approaches examine the use of 3 types of antigen: irradiation-attenuated, live helminths; somatic extracts of helminths; and the metabolic or excretory/secretory (E/S) antigen produced by the *in vitro* culture of helminths. Additional aspects which must be considered include the use of various adjuvants as potentiators of the immune response, the effects of different routes of immunization on the development of protective immunity and the effectiveness of heterologous antigens as immunogens in order to replace homologous antigens where the latter are ineffective or difficult to obtain.

RECENT DEVELOPMENTS IN VACCINE PRODUCTION

Immunization with irradiation-attenuated helminths

The only vaccines against helminths which have been produced on a commercial or widespread scale have been those based on the efficacy of irradiated infective larvae, irradiated to a level at which they are capable of surviving, migrating and producing functional antigens, but at which they have a greatly reduced pathogenicity. Examples of these are vaccines against *Dictyocaulus viviparus* in cattle, *Dictyocaulus filaria* in sheep and *Ancylostoma caninum* in dogs. The activity and efficacy of such vaccines is reviewed by Miller (1971; 1978) and Peacock & Poynter (1980).

Irradiation-attenuated forms have been examined experimentally for their efficacy in immunization against a variety of other helminth infections. Thus, irradiation-attenuated eggs and oncospheres of cestodes have been shown to induce protection against infection with metacestodes in the intermediate host (Urquhart, McIntyre, Mulligan, Jarrett & Sharpe, 1963; Wickerhauser, 1971). However, these irradiated forms have been superseded largely by the experimental use of the E/S antigens of oncospheres of cestodes. Irradiated metacestodes have been used also to immunize against infections with adult cestodes in dogs. Movsesijan, Sokolic & Mladenovic (1968) demonstrated that dogs were protected against a challenge infection of *Echinococcus granulosus* by immunization with brood capsules exposed to X-irradiation. However, Herd, Chappel & Biddell (1975) found this approach to be potentially dangerous since X-irradiated worms regenerated in some animals reaching full maturity and shedding large numbers of infective eggs.

Protection against infection following the administration of irradiation-attenuated larvae has been seen to occur in a variety of nematode infections, including those listed above. For example, it was possible to confer a high degree of protection against infection with *Haemonchus contortus* by immunization of adult sheep with irradiated infective larvae of *H. contortus* (Urquhart, Jarrett, Jennings, McIntyre & Mulligan, 1966). Similarly, irradiated larvae of *Trichostrongylus colubriformis* were highly effective immunogens in 6 to 10-month-old sheep (Gregg & Dineen, 1978). Unfortunately, young lambs cannot be immunized in this way since the immunization of young lambs of up to 5 months of age has proved to be unsuccessful (Benitez-Usher, Armour, Duncan, Urquhart & Gettinby, 1977). Similarly, when 3-month-old lambs were immunized with irradiated larvae of *T. colubriformis*, although some of the lambs ('high responder lambs') developed partial protection, the remainder of the lambs did not respond to the immunizing infections (Gregg, Dineen, Rothwell & Kelly, 1978). This unresponsiveness is of major importance since, in endemic areas, young lambs may be subjected to severe parasitism prior to the time at which they are able to respond with protective immunity. Thus, before immunization against such parasites can be effective, the inability of lambs to initiate a protective immune response must be examined. It is known that young lambs are capable of producing specific antibody in response to infection with *H. contortus* (Valera-Diaz, 1970). In addition, they are able to initiate the self-cure reaction (Chen, 1972). However, cellular immunity, as judged by the ability of peripheral blood lymphocytes to undergo *in vitro* antigen-induced transformation, is lacking until the lambs reach 4–5 months of age (Monsell, Lloyd & Soulsby, unpublished observations). In addition, the inability of young lambs to respond with protective immunity possibly may be associated with the fact that young lambs are unable to mount a secretory IgA immunoglobulin response comparable to that seen in the abomasum of adult sheep (Duncan, Smith & Dargie, 1978). Modulation of these responses would appear to be necessary before successful vaccination against such gastro-intestinal helminths can be achieved.

Irradiation-attenuated forms have been used with success to immunize animals against infection with *Schistosoma* species (see review by Taylor, 1980). Experiments with *Schistosoma mattheei* and *Schistosoma bovis* in sheep and cattle have demonstrated that immunization with irradiated, syringe-transformed, schistosomula is highly effective. Thus, irradiated *S. mattheei* schistosomula, given subcutaneously or intramuscularly to sheep, induced a greater than 60 % reduction in the level of infection seen upon challenge infection with the homologous parasite, even when this challenge infection was delayed for up to 55 weeks after immunization (Bickle, Taylor, James, Nelson, Hussein, Andrews, Dobinson & Marshall, 1979). Similar results were seen following the exposure of sheep and cattle to subcutaneous, intramuscular or percutaneous infection with *S. bovis* (Bushara, Hussein, Saad, Taylor, Dargie, Marshall & Nelson, 1978; Taylor, James, Bickle, Hussein, Andrews, Dobinson & Nelson, 1979). Of particular importance is the fact that calves, immunized both experimentally and in the field, had significantly higher growth rates, superior body composition, decreased faecal and tissue egg counts and lower adult worm counts. In addition, immunized calves showed milder histopathological and haematological changes than did the non-immunized control animals (Hussein, Taylor & Dargie, 1980). These results clearly demonstrate the economic advantages which can be attained through immunization against schistosomiasis in endemic areas despite the fact that, experimentally, the efficacy of the vaccine as judged by adult worm counts was not absolute, reaching only 60–70 %.

Similar use of irradiation-attenuated cercariae and schistosomula to immunize against infection with *Schistosoma mansoni* has produced varied results. Thus, although mice were effectively immunized against infection, the immunization of baboons did not produce significant protection suggesting that immunization of man with irradiation-attenuated forms will not be successful (Taylor, 1980).

Less success has been achieved in immunization against infection with *Fasciola hepatica*. Although attempts to immunize cattle against infection with *F. hepatica* and *Fasciola gigantica* have met with some success, similar attempts in sheep, which show little acquired resistance to infection with *F. hepatica*, have proved disappointing (Bitakaramire, 1973; Nansen, 1975; Campbell, Gregg, Kelly & Dineen, 1978).

Immunization with helminth extracts

Limited and variable success has been demonstrated when animals have been immunized with killed helminths or crude somatic extracts of helminths (see review by Clegg & Smith, 1978). Further, most of the successful experiments have been conducted in laboratory animal models and, in many cases, such immunogens have yet to be examined for their efficacy in domestic animals, primates and man.

Although Miller (1932) and Campbell (1936) were able to demonstrate that rats were partially or completely protected against infection by prior immunization with strobilate or metacestode homogenates of *Taenia taeniaeformis*, killed eggs and oncospheres of *Taenia hydatigena* failed to induce resistance against the homologous parasite in sheep (Gemmell, 1964, 1969). Gemmell therefore concluded that the functional antigens of injected eggs and oncospheres appeared to be released only by viable and metabolizing organisms, since only these were able to stimulate a protective immune response. However, recent studies have repeated the demonstration that functional antigens are present in, and can be extracted from, the somatic tissues of the metacestodes of *T. taeniaeformis* and *Taenia pisiformis* (Heath, 1976; Kwa & Liew, 1977; Lloyd, 1979; Ayuya & Williams, 1979). Further, somatic antigens derived from adult strobilar material of *Taenia saginata* were capable of inducing complete protection against infection with *T. saginata* in cattle (Gallie & Sewell, 1976). On the whole, the immunization of dogs against adult cestodes has been less successful although extracts of protoscolices, hydatid cysts and adult parasites induced some protection against infection with *E. granulosus* (Turner, Berberian & Dennis, 1936; Gemmell, 1962).

Many attempts to induce resistance against nematodes and trematodes by immunization with homogenates and soluble extracts of various stages of the life-cycle have proved to be ineffective or they only marginally increased resistance against infection with the homologous parasite. In addition, the results obtained have been very variable. For instance, Kerr (1938) and Bindseil (1969) were unable to induce immunity in mice with extracts of the larvae or adults of *Ascaris suum*. In contrast, Stromberg & Soulsby (1977*a*) demonstrated a significant level of resistance in guinea pigs immunized against *A. suum* with a soluble extract of adult worms, but not with extracts from the larval stages.

Similar variable results have been seen in experiments designed to induce protection against infection with *Trichinella spiralis*. Thus, Berntzen (1974) was unable to immunize rats against *T. spiralis* using extracts of adults and larvae. However, Despommier & Muller (1976) have isolated α and β granules from cell-free homogenates of the muscle larvae of *T. spiralis*. These granules are found in the stichocytes of the stichosome and have been shown to be highly effective immunogens in rats and mice (Despommier, Campbell & Blair, 1977).

Experimentally, significant progress has been seen also in the evaluation of the functional antigens of *T. colubriformis*. Rothwell & Love (1974) demonstrated that a single injection of a soluble protein extracted from *T. colubriformis* 4th-stage larvae or adults, but not 3rd-stage larvae, stimulated a high level of resistance against challenge infection in guinea pigs. This extract was effective when administered by subcutaneous, intradermal, intraperitoneal and intraduodenal routes (Rothwell, 1978) but the efficacy of the extract needs to be examined in sheep.

Immunization with in vitro*-produced metabolites*

The ability of infections with helminth parasites to stimulate protection against reinfection with the homologous parasite in animals and man; the effectiveness of live, but irradiation-attenuated vaccines; and the ability of hatched, activated oncospheres of cestodes to induce protection against infection with metacestodes when injected intramuscularly into animals, led to the examination of *in vitro*-produced metabolic products as immunizing agents. These secreted antigens are, on the whole, highly effective immunogens and induce protection against reinfection with cestodes and nematodes.

The importance of metabolic substances as immunizing agents was demonstrated by Rickard & Bell (1971*a*). A solid protective immunity against *Taenia ovis* and *T. taeniaeformis* was developed by lambs and rats, respectively, following their exposure to oncospheres grown within millipore chambers implanted into the peritoneal cavity. Additional experiments demonstrated that E/S antigens, collected following the *in vitro* culture of oncospheres, were highly effective in the immunization of sheep against infection with *T. ovis* and *T. hydatigena*, in the immunization of cattle against infection with *T. saginata* and in the immunization of rabbits against infection with *T. pisiformis* (Rickard & Bell, 1971*b*; Heath, 1973; Rickard & Adolph, 1975; Lloyd, 1979). Although Heath (1973) demonstrated that these functional metabolic antigens could only be isolated from metacestodes of *T. pisiformis* less than 15 days of age, other results have demonstrated that these functional metabolic antigens can be produced *in vitro* by the mature metacestodes of *T. taeniaeformis* and *Mesocestoides corti* (Kowalski & Thorson, 1972; Kwa & Liew, 1977; Ayuya & Williams, 1979). It is possible, however, that different mechanisms of protection are associated with the use of E/S antigens obtained from metacestodes of less than 15 days of age as compared with those obtained from mature metacestodes. Thus, when rats were immunized with the E/S antigens of mature metacestodes of *T. taeniaeformis* they were immune to reinfection, but immune serum from these animals could not protect recipient rats against infection with *T. taeniaeformis* (Ayuya & Williams, 1979). This is in contrast with the fact that immunity can be passively transferred with immune serum collected from rats infected with eggs of *T. taeniaeformis* and therefore exposed to the E/S antigens of the young metacestodes (Leid & Williams, 1974) and the fact that a maternal transfer of immunity, presumably associated with serum antibodies, is seen when heifers are immunized intramuscularly with E/S antigens collected during the *in vitro* culture of oncospheres of *T. saginata* (Rickard & Adolph, 1975; Lloyd, 1979).

Metabolic antigens have been examined also for their ability to induce protection against adult cestode infections. Dogs, exposed to the metabolic antigens of *T. pisiformis* by exposure to pre-patent infections, patent infections or repeated doses of eggs of *T. pisiformis* exhibited no acquired immunity to reinfection (Rickard, Coman & Cannon, 1977). On the other hand, secretions from adult *E. granulosus* maintained in culture when administered to dogs increased their resistance to infection with *E. granulosus* and decreased the fecundity of the surviving parasites (Herd *et al.* 1975). However, when this work was extended the results were inconclusive since some of the control dogs were found to exhibit some resistance to infection. The mechanism of the resistance exhibited by these control dogs is unclear. This resistance may be associated with the development of non-specific immunity since these animals were injected with Complete Freund's Adjuvant (CFA) and *Bordetella pertussis* adjuvant. Alternatively, it may be associated with an innate genetic resistance to infection. More information is required before immunization against infection with *E. granulosus* in dogs can be considered feasible (Herd, 1977).

Metabolites secreted *in vitro* have been used to immunize against infection with nematodes. Of these, the secretory antigens associated with moulting appear to be of particular importance. Thus, while results from experiments designed to examine immunization against infection with *A. suum* in mice and guinea pigs have produced variable results (Clegg & Smith, 1978), a recent experiment by Stromberg, Khoury & Soulsby (1977) demonstrated the importance of the moulting fluid produced by *A. suum* during the moult from the 3rd larval stage to the 4th larval stage. These authors examined the efficacy of soluble proteins produced in culture by the 2nd, 3rd and 4th larval stages and adult worms of *A.*

suum, as well as secretions produced during the moult from 3rd stage to 4th stage larvae. Of these, only the moulting fluid induced significant protection against challenge infection. The variability in the results seen in other experiments which examined the immunizing potential of metabolities from various stages of *A. suum* is probably associated with such factors as the use of different culture conditions, the varied protein concentration used for immunization and variations in the immunizing schedule.

Variable results have been seen also when metabolites have been collected *in vitro* from other nematodes. Thus, while Chipman (1957) demonstrated that metabolites collected from adult *T. spiralis* were effective when used to immunize mice, Berntzen (1974) was unable to repeat this observation although he and Campbell (1955) demonstrated that the secretions collected during the *in vitro* growth of muscle larvae were effective immunogens. The antigenic components of these secretions have since been shown to be similar to those contained in the secretory granules of the stichosome (Despommier & Muller, 1976). Other experiments have demonstrated that metabolites from 4th-stage, but not 3rd-stage larvae of *T. colubriformis* were immunogenic in guinea pigs (Rothwell & Love, 1974); metabolites secreted by *H. contortus* during the moult from the 3rd to the 4th larval stage were effective immunogens in sheep (Ozerol & Silverman, 1970); and the *in vitro*-secreted products of adult *Nippostrongylus brasiliensis*, but not *Nematospiroides dubius*, produced protection against homologous infection when used to immunize mice (Day, Howard, Prowse, Chapman & Mitchell, 1979). The varied results seen in such immunization studies performed in experimental animal models means that no conclusions can be made as to the possible efficacy of such antigens in protection against infection with nematodes in domesticated animals and in man.

Despite the successes that have been observed experimentally in immunization against infections with cestodes and nematodes using the metabolic products of the helminths produced during *in vitro* culture, metabolic products of trematodes have not been seen to be highly effective immunogens. Metabolic antigens from cercariae and adults of *S. mansoni* have failed, on the whole, to induce protection against infection when used to immunize mice (Sadun & Lin, 1959; Minard, Murrell & Stirewalt, 1977). In addition, extremely variable results have been obtained when rats and mice were immunized against infection with *F. hepatica* using metabolic antigens. For instance, Lang (1976), Lang & Hall (1977) and Rajasekariah, Mitchell, Chapman & Montague (1979) were able to immunize mice or rats against infection with *F. hepatica* using *in vitro* culture antigens collected from 16-day-old or 4-week-old flukes but not with those collected from mature flukes. In contrast, neither Lehner & Sewell (1979) nor Davies, Rickard, Smyth & Hughes (1979) were able to immunize rabbits, rats and mice with metabolic products of newly excysted or immature *F. hepatica*, although Howell & Sandeman (1979) demonstrated that a complex of rat immunoglobulin and *F. hepatica* metabolic antigen did induce protection against infection. It is possible that complexed antigen is more immunogenic. These metabolic antigens have not been tested for their immunogenicity in sheep which develop little protective immunity against reinfection. Unfortunately, it is likely that they will be ineffective since, although metacercariae administered *via* aberrant routes generated strong resistance to reinfection with *F. hepatica* in the rat, the inability of sheep to develop homologous protection against infection was not overcome (Kelly & Campbell, 1979).

Isolation and characterization of functional antigens

A few functional antigens have now been isolated and partially characterized. However, the isolation of E/S antigens has been complicated by the presence in the media of complex constituents such as serum. The development of defined low molecular weight culture media, such as that described by Stromberg *et al.* (1977) for the culture of larvae of *A. suum*, will aid greatly in the isolation and physico-chemical characterization of functional antigens. In addition, in some cases, it has been demonstrated that the E/S antigens produced by helminths can be readily extracted from homogenates of the parasite. Thus, it is possible that in the future, many functional antigens will be isolated from either source.

The exogenous antigens secreted in culture by the developing oncospheres of *T. pisiformis* have been examined by Heath (1973). Their immunogenicity was completely destroyed by oxidation which affected the levels of cysteine and methionine. Six protein components were revealed on electrophoresis and, of the 6 antigenic fractions seen on immunological analysis, only 2 were distinct for secretions. The remaining bands cross-reacted with antigens seen in a somatic extract of the developing metacestode.

In addition, a functional antigen has been isolated and partially characterized from metacestodes of *T. taeniaeformis* (Kwa & Liew, 1977). Fractionation of a somatic extract and the E/S antigens of mature metacestodes of *T. taeniaeformis* revealed that a highly antigenic protein with a molecular weight of 140000 Daltons could be obtained from both sources. Since this protein comprised 6·8–8·1 % of the original starting material, Kwa and Liew (1977) concluded that it is more appropriate to isolate such a functional antigen from somatic material rather than from the E/S antigens.

Another protective antigen which has been isolated and partially characterized is that produced *in vitro* during the moult of *A. suum* 3rd stage larvae to the 4th stage (Stromberg, 1979). A single protein was seen when this antigen, the ACF antigen, was subjected to double immunodiffusion in gel and cellulose acetate electrophoresis. The ACF antigen, which may be a polymer, is estimated to contain 79 % protein and 22 % carbohydrate.

The functional antigens produced by larvae of *T. spiralis* in culture have been isolated by sucrose gradient centrifugation of cell-free homogenates of the muscle larvae of *T. spiralis* (Despommier & Muller, 1976). Two types of granules, α and β granules, were identified and at least 4 unique antigens have been attributed to each of these granule types. The antigens in the granules were found to be identical to those secreted by the larvae in culture since there was complete cross-reactivity between them.

Non-specific immunization

Adjuvants have been shown to potentiate the immune response and to non-specifically increase protection against infection with a variety of helminths. Their mode of action, however, is not known although, in some cases, the increased protective immune response has been associated with the non-specific activation of macrophages.

Some examples of the non-specific activity of adjuvants are that mature metacestodes of *E. granulosus* and *T. taeniaeformis* disintegrated and degenerated when transferred to rats immunized with CFA (Valera-Diaz, Williams, Coltorti & Williams, 1974; Musoke & Williams, 1978). Similarly, dogs immunized with CFA and *B. pertussis* appeared to show some degree of protection against infection with *E. granulosus* (Herd, 1977). Also, Bacillus Calmette-Guerin (BCG) has been shown to increase resistance against infection with *Echinococcus multilocularis* in cotton rats and *E. granulosus* in gerbils. This increased resistance would appear to be associated with activated macrophages since BCG treatment activated macrophages to kill protoscolices *in vitro* (Thompson, 1976; Reuben & Tanner, 1979). In addition, the administration of BCG increased protection against infection with *S. mansoni* in mice. However, very high doses of BCG ($3{\cdot}75 \times 10^7$ colony-forming units), which produced a disseminated granulomatous response were required and when Rhesus monkeys were given 10^7 CFU in each thigh, protection against infection with *S. mansoni* was not induced (Maddison, Chandler & Kagan, 1978). Generally, adjuvants have been used close to the time of infection. Further studies on the long-term potential of adjuvants are required.

The use of adjuvants as a vehicle for antigen in order to potentiate the immune response has produced very varied results dependent on the type of adjuvant used and the species of animal immunized. For instance, CFA was an effective adjuvant when albino rats and CF1 mice were immunized against infection with *T. taeniaeformis* (Kwa & Liew, 1977; Lloyd, 1979). In addition, when CFA and Incomplete Freund's Adjuvant were used as a vehicle for the E/S antigens of oncospheres of cestodes, immunized sheep and cattle were protected against infection with the metacestodes (Rickard & Bell, 1971*b*; Lloyd, 1979). Conversely, Ayuya & Williams (1979) found that when Spartan (Spb(SD)BR) rats were immunized with antigen emulsified in CFA these rats did not develop resistance to infection with *T. taeniaeformis*. In fact, rats immunized with CFA and antigen or CFA alone contained higher numbers of parasites than did the untreated control rats. In contrast, in the same experiment, *B. pertussis* was highly effective as an adjuvant, as was aluminium hydroxide. The route of immunization was important also. For instance, rats immunized orally with antigen, and those given antigen *per os* plus *B. pertussis* intramuscularly, had similar levels of resistance to infection with *T. taeniaeformis*. However, the level of protection against infection was greater when the antigen was administered with *B. pertussis via* the intramuscular route (Ayuya & Williams, 1979).

A difference in the efficacy of various adjuvants was seen also when guinea pigs were immunized against infection with *T. colubriformis*. Administration of the antigen with levamisole, *B. pertussis*, CFA and IFA did not increase its efficacy, but the use of aluminium hydroxide did increase the degree of resistance seen on challenge infection (Rothwell, 1978). The antigen was effective when given subcutaneously, intraperitoneally, intradermally and intraduodenally, but not orally. It is probable that this antigen, when administered orally, was degraded by gastric juices.

It is evident that extensive studies are required in immunization trials to examine the efficacy of a variety of adjuvants as well as to examine the route of immunization.

Heterologous immunization

Immunization has been studied primarily with antigens from the homologous parasite although there is considerable evidence that cross-protection is important in many of the helminth infections of man and domesticated animals (Nelson, 1974). Since, from the practical standpoint, antigens from the homologous parasite cannot be used on many occasions, the effectiveness of heterologous antigens must be evaluated. Immunization with heterologous antigens has given variable results but, in some instances, a marked cross-protection may be demonstrated.

Gemmell and his colleagues demonstrated a significant degree of cross-protection in sheep immunized intramuscularly with hatched, activated oncospheres of closely related cestodes, but not with oncospheres from cestodes which parasitize other species of animals (Gemmell & Johnstone, 1977). Gemmell concluded that the relationships of the intermediate hosts of the cestodes was a more important determinant in heterologous immunity than the systemic relationship between the species of cestodes and that there were some similar antigens in the oncospheres of taeniids that parasitize sheep. However, little cross-protection was seen when sheep were immunized with *T. ovis* or *T. hydatigena* and challenged with *E. granulosus* (Gemmell, 1966). Similarly, when Heath, Lawrence & Young (1979) hyperimmunized sheep with *E. granulosus*, *T. hydatigena* or *T. ovis* and challenged them with the heterologous species only 37–53 % protection against infection was seen. Heath *et al.* (1979) considered this level of resistance to be too low to be of epidemiological significance. Other studies by Wickerhauser, Zukovic & Dzakula (1971) and Rickard & Adolph (1975) demonstrated that immunization of cattle with oncospheres or E/S antigens of *T. hydatigena* increased their resistance to infection with *T. saginata* but the effectiveness of *T. hydatigena* as a heterologous antigen in this situation has not always been confirmed.

However, a very high (96 %) level of protection against infection with *T. taeniaeformis* in mice was seen following prior immunization with the E/S antigens of oncospheres of *T. saginata*. Immunization with a somatic antigen and prior infection with viable metacestodes of *T. crassiceps* was effective also (Lloyd, 1979). Furthermore, the converse was also true. Thus, the E/S products of the heterologous parasite, *T. taeniaeformis*, induced a 63 % level of protection in calves against infection with *T. saginata*. In addition, these E/S antigens induced a colostral transfer of immunity to neonatal calves when the antigens were given, either intramuscularly or *via* the intramammary route, to pregnant heifers (Lloyd, 1979). The effectiveness of this heterologous antigen in the stimulation of protective immunity against infection with *T. saginata* in cattle is of particular importance since this would avoid the need for human donors of immunizing material.

Heterologous immunization against nematode infections has frequently been ineffective, but heterologous immunity has been demonstrated on some occasions. For instance, Crandall, Crandall & Arean (1967) demonstrated an increased resistance to *A. suum* in mice induced by infection with *Nippostrongylus brasiliensis*; rats infected with *T. spiralis* were protected against infection with *Strongyloides ratti* (Kazacos, 1976); and guinea pigs, injected *via* the mesenteric vein with eggs or larvae of *Toxocara canis* or larvae of *Ancylostoma caninum*, were protected against infection with *A. suum* (Stromberg & Soulsby, 1977*b*).

Recent studies by Dineen, Gregg, Windon, Donald & Kelly (1977) clearly demonstrated that immunization of sheep with irradiated larvae of *T. colubriformis* induced a marked, 81 %, level of protection against homologous challenge while the levels of resistance upon heterologous challenge with *Trichostrongylus vitrinus* and *Nematodirus spathiger* were 34 and 14 %, respectively. This indicated the presence of some degree of antigen-sharing between the closely related parasites, *T. colubriformis* and *T. vitrinus*, but not the unrelated parasite, *N. spathiger*. However, challenge infection of similarly immunized sheep with mixed infections of *T. colubriformis*, *T. vitrinus* and *N. spathiger* resulted in a 98–100 % level of protection against infection with all 3 parasites. This demonstrated that although antigen recognition appeared specific, the effector mechanism of expulsion was not. These results have important implications for immunization against gastro-intestinal helminth infections in the field since, although naturally infected animals are infected with a variety of helminths, the use of heterologous immunization is likely to be successful.

Homologous immunization of sheep against *F. hepatica* has not been successful, but the above results suggested that the infection of sheep with *T. hydatigena* might specifically or non-specifically potentiate resistance to *F. hepatica* since both parasites affected the liver and *T. hydatigena* induces a high degree of homologous resistance. Indeed, sheep infected with *T. hydatigena* were resistant to a challenge infection with *F. hepatica* given 12 weeks or 9 months, but not 3 weeks, after primary infection with *T. hydatigena* (Campbell, Kelly, Townsend & Dineen, 1977; Dineen, Kelly & Campbell, 1978). In contrast, Hughes, Harness & Doy (1978) were unable to demonstrate the development of cross-protection between *T. hydatigena* and *F. hepatica* in sheep, cattle and goats. The differences in the above results might be associated with the fact that different breeds of sheep were used in the two studies, since differences in resistance to infection with *H. contortus* have been observed in different breeds of sheep (Preston & Allenby, 1978) and marked genetic variation in resistance to infection with nematodes has been seen in various strains of mice (Mitchell, 1979). As genetic resistance to disease is becoming of increasing importance in livestock development programmes, the effects of genetic resistance must be considered in vaccination trials.

Another parasite against which active immunization using homologous parasite antigens is generally ineffective is *S. mansoni*. However, Nelson, Amin, Saoud & Teesdale (1968) demonstrated that a single exposure to cercariae of either *S. bovis*, *S. mattheei* or *S. rodhaini* conferred a high degree of heterologous immunity against a challenge infection with *S. mansoni* in mice. In addition, cross-protection following infection with the heterologous parasite has been demonstrated between *S. haematobium* and *S. mansoni* in hamsters (Smith, Clegg & Webbe, 1976), *S. mansoni* and *F. hepatica* in mice (Christensen, Nansen, Frandsen, Bjornenoe & Monrad, 1978), and baboons immunized with repeated exposures to *S. rodhaini* and *S. bovis* developed resistance to *S. mansoni* (Taylor, Nelson, Smith & Andrews, 1973). Further, mice and hamsters, immunized with antigenic extracts of *F. hepatica*, developed a significant reduction in their worm burdens when challenged with *S. mansoni* (Hillyer, del Llano de Diaz & Reyes, 1977). Conversely, irradiated schistosomula of *S. mansoni* were ineffective when used to immunize sheep against infection with *S. mattheei* and immunization with *S. mattheei* did not induce a high level of protection against *S. bovis*. These results suggest that species-specific immunization with irradiated schistosomula will be necessary to control schistosomiasis in domestic animals (Taylor, 1980). However, this premise is not necessarily true for the control of human schistosomiasis.

Oral immunization

The role of the intestine as a first line of defence against infection has led to an investigation of oral immunization. Thus, the oral presentation of antigens is now well established in vaccination against polio in man and has been found beneficial against the effects of infection with *Escherichia coli* in pigs (Porter, Kenworthy & Allen, 1974). In addition, Frerichs (1976) has reported that 20% of veterinary vaccines effective against bacteria and viruses which have been introduced into the United Kingdom are given by a non-parenteral, primarily oral route. In contrast, only limited work has been performed on the use of oral immunization against helminth infections.

Oral presentation of bacterial and viral vaccines results in a local stimulation of the secretory immunologic system of the gastro-intestinal tract and the production of IgA immunoglobulins. These intestinal IgA immunoglobulins are associated with protection against a number of bacterial and viral infections and IgA antibodies mediate this protection by means of viral neutralization, bacterial agglutination, and through the inhibition of bacterial mobility and bacterial adhesion to epithelial cells (Abou-Youssef & Ristic, 1975; Bohl & Saif, 1975; Fubara & Freter, 1973; Porter, Parry & Allen, 1977).

Since the portal of entry and site of entrance of many helminth infections is the gastro-intestinal tract, the role of the intestine in protection against infection with helminths needs further examination. However, only recently has the role of the local secretory IgA response of the gastro-intestinal tract been evaluated in intestinal parasitism. Elevated levels of IgA antibodies have been demonstrated in a number of helminth infections. These include *T. colubriformis* and *H. contortus* in sheep, *N. brasiliensis* in rats and *A. suum* in pigs (Smith, 1977; Sinski & Holmes, 1977; Cripps & Adams, 1978; Rhodes, McCullough, Mebus & Klucas, 1978). In addition, it has been suggested that IgA immunoglobulins are associated with protection against infection with *T. spiralis* in the rat (Despommier, McGregor, Crum & Carter, 1977). The conclusive evidence associating these elevated levels of IgA immunoglobulins with protection against infection with helminths is often lacking and the intestinal immune response may be associated also with other classes of immunoglobulins, thymus-dependent lymphocytes and non-lymphoid effector cells (Wakelin, 1978).

However, recently it was demonstrated that the infection of mice with *T. taeniaeformis* resulted in the production of protective intestinal IgA antibodies (Lloyd & Soulsby, 1978). For this reason, oral immunization against infection with *T. taeniaeformis* was examined. Indeed, this route of immunization resulted in a highly effective protective immune response against infection with *T. taeniaeformis*

(Ayuya & Williams, 1979; Lloyd, 1979). Somatic antigens extracted from the metacestodes of *T. taeniaeformis* as well as *in vitro* produced excretory/secretory antigens collected from oncospheres and metacestodes were highly effective immunogens when administered orally to mice and rats. Also, laboratory animals were protected against infection with *T. spiralis* following the oral administration of antigens obtained from the muscle larvae (Vernes, 1976).

Further studies are needed to examine the oral route of administration of helminth antigens in immunization against a variety of helminth infections in domestic animals and man. This route of immunization would be particularly attractive since the oral presentation of antigen may result, in some species of animals, in both a local intestinal immune response and a concommitant colostral transfer of immunity to the newborn animal (Lloyd, 1980). However, care must be taken when examining the presentation of antigens *via* the oral route, since the presentation of antigens to the gastro-intestinal tract has been shown to induce systemic tolerance (Thomas & Parrot, 1974) despite the fact that there may be a concomitant intestinal IgA response and a local cell-mediated immune response (Muller-Schoop & Good, 1975; Tomasi, 1976). Systemic tolerance should not be of importance in immunization against helminth infections whose life-cycle is restricted to the gastro-intestinal tract, but it could be of major importance in the control of those helminth infections which undergo tracheal or somatic migration within the host. However, systemic tolerance was not evident when rats were orally immunized against infection with *T. taeniaeformis* since such rats were resistant to infection by both the oral and parenteral routes (Ayuya & Williams, 1979). However, it would appear that different mechanisms of immunity are seen after oral immunization, since serum from orally immunized rats could not be used to passively immunize recipient animals against infection with *T. taeniaeformis* (Ayuya & Williams, 1979). This is in contrast to the fact that serum from rats orally infected with eggs of *T. taeniaeformis* did induce a passive transfer of immunity (Leid & Williams, 1974). It is possible that oral immunization stimulated a local intestinal IgA response. This hypothesis is not ruled out by the fact that such orally immunized rats were protected against parenteral challenge infections since the metacestodes of *T. taeniaeformis* develop in the liver and IgA antibodies which are removed from the circulation by the liver (Orlans, Peppard, Reynolds & Hall, 1978) could affect the developing oncospheres within the liver. Further, IgA immunoglobulins of bile origin have been shown to have an inhibitory effect on *in vitro* larval production by *T. spiralis* (Jacqueline, 1980). The advantage of local stimulation of the secretory immunologic system of the intestine is that cells specific for such IgA antibodies will be transferred to the common muscosal immunologic system of the mammary gland and result in a maternal transfer of immunity to the suckling neonate (Lloyd, 1980).

CONCLUSIONS

From the literature discussed in this review it is obvious that the production of new vaccines against helminths is a viable possibility. This is particularly so in the field of veterinary medicine, and the immunological control of infections such as bovine cysticercosis, as well as being of considerable economic advantage to the farmer, would have important public health implications. However, the experiments described do demonstrate the complexity of the problems involved. Only in a few instances have antigens been isolated and at least partially characterized. Isolation and characterization of functional antigens is necessary since, in the absence of effective *in vitro* culture systems to produce large quantities of helminth material, it is possible that defined antigens could be produced by means of genetic engineering. Further, from a practical standpoint, heterologous immunization is required in some instances to circumvent the need for large quantities of material of human origin and the need for parasites from which large quantities of material are difficult to obtain. Thus, the progress in the development of heterologous immunization is encouraging. Although the percentage protection produced by antigens from heterologous parasites has not normally been as great as that induced by those from the homologous parasite, it is possible that manipulation of the immunizing schedule or the isolation of purified functional antigens could result in a highly effective protective immune response. Such factors as the route of immunization (i.e. oral versus parenteral) and the efficacy of a variety of adjuvants must be considered in all studies examining immunization against infection with helminths. In addition, the species and breed of animal in which the antigen is tested is important, since the use of animals with an innate genetic resistance to the infection would bias the experiment in favour of the antigen. However, the use of such genetic 'responder' and 'non-responder' animals may assist in the elucidation of the immune affector and effector mechanisms in helminth infections.

REFERENCES

Abou-Youssef. M. H. & Ristic, M. (1975). Protective effect of immunoglobulins in serum and milk of sows exposed to transmissible gastro-enteritis virus. *Canadian Journal of Comparative Medicine* **39**, 41–5.

Ayuya, J. M. & Williams, J. F. (1979). The immunological response of the rat to infection with *Taenia taeniaeformis*. VII. Immunization by oral and parenteral administration of antigens. *Immunology* **36**, 825–34.

Benitez-Usher, C., Armour, J., Duncan, J. L., Urquhart, G. M. & Gettinby, G. (1977). A study of some factors influencing the immunization of sheep against *Haemonchus contortus* using attenuated larvae. *Veterinary Parasitology* **3**, 327–42.

Berntzen, A. K. (1974). Effects of environment on the growth and development of *Trichinella spiralis in vitro*. In *Trichinellosis. Proceedings of the 3rd International Conference in Trichinellosis*, (ed. C. W. Kim), pp. 25–30. New York: Intext Educational Publishers.

Bickle, Q. D., Taylor, M. G., James, E. R., Nelson, G. S., Hussein, M. F., Andrews, B. J., Dobinson, A. R. & Marshall, T. F. de C. (1979). Further observations on immunization of sheep against *Schistosoma mattheei* and *S. bovis* using irradiation-attenuated schistosomula of homologous and heterologous species. *Parasitology* **78**, 185–93.

Bindseil, E. (1969). Immunity to *Ascaris suum*-I. Immunity induced in mice by means of material from adult worms. *Acta Pathologica et Microbiologica, Scandanavica* **77**, 223–34.

Bitakaramire, P. K. (1973). Preliminary studies on the immunization of cattle against fascioliasis using gamma-irradiated metacercariae of *Fasciola gigantica*. In *Isotopes and Radiation in Parasitology III*, pp. 23–32. Vienna: International Atomic Energy Association.

Bohl, E. H. & Saif, L. J. (1975). Passive immunity in transmissible gastro-enteristis of swine: Immunoglobulin characteristics of antibodies in milk after innoculating virus by different routes. *Infection and Immunity* **11**, 23–32.

Bushara, H. O., Hussein, M. F., Saad, A. M., Taylor, M. G., Dargie, J. D., Marshall, T. F. de C. & Nelson, G. S. (1978). Immunization of calves against *Schistosomà bovis* using irradiated cercariae or schistosomula of *S. bovis*. *Parasitology* **77**, 303–11.

Campbell, C. H. (1955). The antigenic role of the excretions and secretions of *Trichinella spiralis* in the production of immunity in mice. *Journal of Parasitology* **41**, 483–91.

Campbell, D. H. (1936). Active immunization of albino rats with protein fractions from *Taenia taeniaeformis* and its larval form *Cysticercus fasciolaris*. *American Journal of Hygiene* **23**, 104–13.

Campbell, N. J., Gregg, P., Kelly, J. D. & Dineen, J. K. (1978). Failure to induce homologous immunity to *Fasciola hepatica* in sheep vaccinated with irradiated metacercariae. *Veterinary Parasitology* **46**, 113–20.

Campbell, N. J., Kelly, J. D., Townsend, R. B. & Dineen, J. K. (1977). The stimulation of resistance in sheep to *Fasciola hepatica* by infection with *Cysticercus tenuicollis*. *International Journal for Parasitology* **7**, 347–51.

Chen, P. M. (1972). Peripheral lymphoid cell response of sheep to gastro-intestinal parasitism. Ph.D. thesis, University of Pennsylvania.

Chipman, P. B. (1957). The antigenic role of the excretions and secretions of adult *Trichinella spiralis* in the production of immunity in mice. *Journal of Parasitology* **43**, 593–8.

Christensen, N. O., Nansen, P., Frandsen, F., Bjornenoe, A. & Monrad, J. (1978). *Schistosoma mansoni* and *Fasciola hepatica*: Cross-resistance in mice. *Experimental Parasitology* **46**, 113–20.

Clegg, J. A. & Smith, M. A. (1978). Prospects for the development of dead vaccines against helminths. *Advances in Parasitology* **16**, 165–218.

Crandall, C. A., Crandall, R. B. & Arean, V. M. (1967). Increased resistance in mice to larval *Ascaris suum* infection induced by *Nippostrongylus brasiliensis*. *Journal of Parasitology* **53**, 214–15.

Cripps, A. W. & Adams, D. B. (1978). Flow and protein composition of intestinal lympth in sheep infected with the enteric nematode *Trichostrongylus colubriformis*. *Australian Journal of Experimental Biology and Medical Science* **56**, 225–35.

Davies, C., Rickard, M. D., Smyth, J. D. & Hughes, D. L. (1979). Attempts to immunize rats against infection with *Fasciola hepatica* using *in vitro* culture antigens from newly excysted metacercariae. *Research in Veterinary Science* **26**, 259–60.

Day, K. P., Howard, R. J., Prowse, S. J., Chapman, C. B. & Mitchell, G. F. (1979). Studies on chronic versus transient intestinal nematode infections in mice. I. A comparison of response to excretory/secretory (ES) products of *Nippostrongylus brasiliensis* and *Nematospiroides dubius* worms. *Parasite Immunology* **1**, 217–39.

Despommier, D. D., Campbell, W. C. & Blair, L. S. (1977). The *in vivo* and *in vitro* analysis of immunity to *Trichinella spiralis* in mice and rats. *Parasitology* **74**, 109–19.

Despommier, D. D., McGregor, D. D., Crum, E. D. & Carter, P. B. (1977). Immunity to *Trichinella spiralis*. II. Expression of immunity against adult worms. *Immunology* **33**, 797–805.

Despommier, D. D. & Muller, M. (1976). The stichosome and its secretion granules in the mature muscle larvae of *Trichinella spiralis*. *Journal of Parasitology* **62**, 775–85.

Dineen, J. K., Gregg, P., Windon, R. G., Donald, A. D. & Kelly, J. D. (1977). The role of immunologically specific and non-specific components of resistance in cross-protection to intestinal nematodes. *International Journal for Parasitology* **7**, 211–15.

Dineen, J. K., Kelly, J. D. & Campbell, N. J. (1978). Further observations on the nature and characteristics of cross-protection against *Fasciola hepatica* produced in sheep by infection with *Cysticercus tenuicollis*. *International Journal for Parasitology* **8**, 173–6.

Duncan, J. L., Smith, W. D. & Dargie, J. D. (1978). Possible relationship of levels of mucosal IgA and serum IgG to immune unresponsiveness of lambs to *Haemonchus contortus*. *Veterinary Parasitology* **4**, 21–7.

Frerichs, G. N. (1976). Non-paranteral vaccines for veterinary use in the United Kingdom. *Development of Biological Standards* **33**, 29–32.

Fubara, E. S. & Freter, R. (1973). Protection against enteric bacterial infection by secretory IgA antibodies. *Journal of Immunology* **111**, 395–403.

Gallie, G. S. & Sewell, M. M. H. (1976). Experimental immunization of 6-month-old calves against infection with the cysticercus stage of *Taenia saginata*. *Tropical Animal Health and Production* **8**, 233–42.

Gemmell, M. A. (1962). Natural and acquired immunity factors interfering with development during the rapid growth phase of *Echinococcus granulosus* in dogs. *Immunology* **5**, 496–503.

Gemmell, M. A. (1964). Immunological responses of the mammalian host against tapeworm infections. I. Species specificity of hexacanth embryos in protecting sheep against *Taenia hydatigena*. *Immunology* **7**, 489–99.

GEMMELL, M. A. (1966). Immunological responses of the mammalian host against tapeworm infection. IV. Species specificity of hexacanth embryos in protecting sheep against *Echinococcus granulosus*. *Immunology* **11**, 325–35.

GEMMELL, M. A. (1969). Immunization of sheep against *Taenia hydatigena* and *T. ovis* with chemically and physically treated embryos. *Experimental Parasitology* **26**, 58–66.

GEMMELL, M. A. & JOHNSTONE, P. D. (1977). Experimental epidemiology of hydatidosis and cysticercosis. *Advances in Parasitology* **15**, 311–69.

GREGG, P. & DINEEN, J. K. (1978). The response of sheep vaccinated with irradiated *Trichostrongylus colubriformis* larvae to impulse and sequential challenge with normal larvae. *Veterinary Parasitology* **4**, 49–53.

GREGG, P., DINEEN, J. K., ROTHWELL, T. L. W. & KELLY, J. D. (1978). The effect of age on the response of sheep to vaccination with irradiated *Trichostrongylus colubriformis* larvae. *Veterinary Parasitology* **4**, 35–48.

HEATH, D. D. (1973). Resistance to *Taenia pisiformis* larvae in rabbits. I. Examination of the antigenically protective phase of larval development. *International Journal for Parasitology* **3**, 485–9.

HEATH, D. D. (1976). Resistance to *Taenia pisiformis* larvae in rabbits: Immunization against infection using non-living antigens from *in vitro* culture. *International Journal for Parasitology* **6**, 19–24.

HEATH, D. D., LAWRENCE, S. B. & YONG, W. K. (1979). Cross-protection between the cysts of *Echinococcus granulosus*, *Taenia hydatigena* and *T. ovis* in lambs. *Research in Veterinary Science* **27**, 210–12.

HERD, R. P. (1977). Resistance of dogs to *Echinococcus granulosus*. *International Journal for Parasitology* **7**, 135–8.

HERD, R. P., CHAPPEL, R. J. & BIDDELL, D. (1975). Immunization of dogs against *Echinococcus granulosus* using worm secretory antigens. *International Journal for Parasitology* **5**, 395–9.

HILLYER, G. V., DEL LLANO DE DIAZ, A. & REYES, C. N. (1977). *Schistosoma mansoni*: acquired immunity in mice and hamsters using antigens of *Fasciola hepatica*. *Experimental Parasitology* **42**, 348–55.

HOWELL, M. J. & SANDEMAN, R. M. (1979). *Fasciola hepatica*: some properties of a precipitate which forms when metacercariae are cultured in immune rat serum. *International Journal for Parasitology* **9**, 41–5.

HUGHES, D. L., HARNESS, E. & DOY, T. G. (1978). Failure to demonstrate resistance in goats, sheep and cattle to *Fasciola hepatica* after infection with *Cysticercus tenuicollis*. *Research in Veterinary Science* **25**, 356–9.

HUSSEIN, M. F., TAYLOR, M. G. & DARGIE, J. D. (1980). Pathogenesis and immunology of ruminant schistosomiasis in the Sudan. In *Isotopes and Radiation in Parasitology*. Vienna: International Atomic Energy Agency (in the Press).

JACQUELINE, J. E. (1980). *Trichinella spiralis*: study of the effects of bile secretory immune factors. In *Third European Multicolloquium of Parasitology*, p. 43. (Abstract).

KAZACOS, K. R. (1976). Increased resistance in the rat to *Strongyloides ratti* following immunization with *Trichinella spiralis*. *Journal of Parasitology* **62**, 493–4.

KELLY, J. D. & CAMPBELL, N. J. (1979). The effect of route of infection on acquired resistance to *F. hepatica* in the rat and sheep. *Research in Veterinary Science* **27**, 205–9.

KERR, K. B. (1938). Attempts to induce an artifical immunity against the dog hookworm, *A. caninum*, and the pig *Ascaris*, *A. lumbricoides suum*. *American Journal of Hygiene* **27**, 52–9.

KOWALSKI, J. C. & THORSON, R. E. (1972). Immunization of laboratory mice against *Tetrathyridia* of *Mesocestoides corti* (Cestoda) using a secretory and excretory antigen and a soluble somatic antigen. *Journal of Parasitology* **58**, 732–4.

KWA, B. H. & LIEW, F. Y. (1977). Immunity in taeniasis–cysticercosis. I. Vaccination against *Taenia taeniaeformis* in rats using purified antigens. *Journal of Experimental Medicine* **146**, 118–31.

LANG, B. Z. (1976). Host–parasite relationships of *Fasciola hepatica* in the white mouse. VII. Effects of anti-worm incubate sera on transferred worms and successful vaccination with a crude incubate antigen. *Journal of Parasitology* **62**, 232–6.

LANG, B. Z. & HALL, R. F. (1977). Host–parasite relationship of *Fasciola hepatica* in the white mouse. VIII. Successful vaccination with culture incubate antigens and antigens from sonic disruption of immature worms. *Journal of Parasitology* **63**, 1046–9.

LEHNER, R. P. & SEWELL, M. M. H. (1979). Attempted immunization of laboratory animals with metabolic antigens of *Fasciola hepatica*. *Veterinary Science Communications* **2**, 337–40.
LEID, R. W. & WILLIAMS, J. F. (1974). Immunological response of the rat to infection with *Taenia taeniaeformis*. I. Immunoglobulin classes involved in passive transfer of resistance. *Immunology* **27**, 195–208.
LLOYD, S. (1979). Homologous and heterologous immunization against the metacestodes of *Taenia saginata* and *Taenia taeniaeformis* in cattle and mice. *Zeitschrift für Parasitenkunde* **60**, 87–96.
LLOYD, S. (1980). Local immune mechanisms against parasites. In *Isotopes and Radiation in Parasitology*. Vienna: International Atomic Energy Agency (in the Press).
LLOYD, S. & SOULSBY, E. J. L. (1978). The role of IgA immunoglobulins in the passive transfer of protection to *Taenia taeniaeformis* in the mouse. *Immunology* **34**, 939–45.
MADDISON, S. E., CHANDLER, F. W. & KAGAN, I. G. (1978). The effect of pretreatment with BCG on infection with *Schistosoma mansoni* in mice and monkeys. *Journal of the Reticuloendothelial Society* **24**, 615–28.
MILLER, H. M. (1932). Acquired immunity against a metazoan parasite by use of non-specific worm materials. *Proceedings of the Society for Experimental Biology and Medicine* **29**, 1125–6.
MILLER, T. A. (1971). Vaccination against the canine hookworm diseases. *Advances in Parasitology* **9**, 153–83.
MILLER, T. A. (1978). Industrial development and field use of the canine hookworm vaccine. *Advances in Parasitology* **16**, 333–42.
MINARD, P., MURRELL, K. D. & STIREWALT, M. A. (1977). Proteolytic, antigenic and immunologic properties of *Schistosoma mansoni* cercarial secretions. *American Journal of Tropical Medicine and Hygiene* **26**, 491–9.
MITCHELL, G. F. (1979). Responses to infection with metazoan and protozoan parasites in mice. *Advances in Immunology* **28**, 451–511.
MOVSESIJAN, M., SOKOLIC, A. & MLADENOVIC, Z. (1968). Studies on the immunological potentiality of irradiated *Echinococcus granulosus* forms: experiments in dogs. *British Veterinary Journal* **124**, 425–32.
MULLER-SCHOOP, J. W. & GOOD, R. A. (1975). Functional studies of Peyer's patches: evidence for their participation in intestinal immune response. *Journal of Immunology* **114**, 1757–60.
MUSOKE, A. J. & WILLIAMS, J. F. (1976). Immunological response of the rat to infection with *Taenia taeniaeformis*: protective antibody response to implanted parasites. *International Journal for Parasitology* **6**, 265–9.
NANSEN, P. (1975). Resistance in cattle to *Fasciola hepatica* induced by X-ray attenuated larvae: results from a controlled field trial. *Research in Veterinary Science* **19**, 278–83.
NELSON, G. S. (1974). Zooprophylaxis with special reference to schistosomiasis and filariasis. In *Parasitic Zoonoses* (ed. E. J. L. Soulsby), pp. 273–85. New York: Academic Press.
NELSON, G. S., AMIN, M. A., SAOUD, M. F. A. & TEESDALE, C. (1968). Studies on heterologous immunity in schistosomiasis. I. Heterologous schistosome immunity in mice. *Bulletin of the World Health Organization* **38**, 9–17.
ORLANS, E., PEPPARD, J., REYNOLDS, J. & HALL, J. (1978). Rapid active transport of immunoglobulin A from blood to bile. *Journal of Experimental Medicine* **147**, 588–92.
OZEROL, N. H. & SILVERMAN, P. H. (1970). Further characterization of active metabolites from histotropic larvae of *Haemonchus contortus* cultured *in vitro*. *Journal of Parasitology* **56**, 1199–205.
PEACOCK, R. & POYNTER, D. (1980). Field experience with a bovine lungworm vaccine. In *Vaccines against Parasites*, (ed. A. E. R. Taylor and R. Muller), pp. 141–148. London: Blackwell Scientific Publications.
PORTER, P., KENWORTHY, R. & ALLEN, W. D. (1974). Effect of oral immunization with *E. coli* antigens on post weaning enteric infection in the young pig. *Veterinary Record* **95**, 99–104.
PORTER, P., PARRY, S. H. & ALLEN, W. D. (1977). Significance of immune mechanisms in relation to enteric infections of the gastrointestinal tract in animals. In *Immunology of the Gut. Ciba Fdn Symp.* **46**, pp. 55–75. Amsterdam: Elsevier/Exerpta Medica/North Holland.
PRESTON, J. M. & ALLONBY, E. W. (1978). The influence of breed in the susceptibility of sheep and goats to a single infection with *Haemonchus contortus*. *Veterinary Record* **103**, 509–12.
RAJASEKARIAH, G. R., MITCHELL, G. F., CHAPMAN, C. B. & MONTAGUE, P. E. (1979). *Fasciola hepatica*: attempts to induce protection against infection by injection of excretory/secretory products of immature worms. *Parasitology* **79**, 393–400.

REUBEN, J. M. & TANNER, C. E. (1979). Immunoprophylaxis with BCG of experimental *Echinococcus multilocularis* infections. *Australian Veterinary Journal* **55**, 105–8.

RHODES, M. B., McCULLOUGH, R. A., MEBUS, C. A. & KLUCAS, C. A. (1978). *Ascaris suum*: specific antibodies in isolated intestinal loop washings from immunized swine. *Experimental Parasitology* **45**, 255–62.

RICKARD, M. D. & ADOLF, A. J. (1975). Vaccination of calves against *Taenia saginata* using a 'parasite-free' vaccine. *Veterinary Parasitology* **1**, 389–92.

RICKARD, M. D. & BELL, K. J. (1971*a*). Immunity produced against *Taenia ovis* and *T. taeniaeformis* infection in lambs and rats following *in vivo* growth of their larvae in filtration membrane diffusion chambers. *Journal of Parasitology* **57**, 571–5.

RICKARD, M. D. & BELL, K. J. (1971*b*). Successful vaccination of lambs against infection with *Taenia ovis* using antigens produced during *in vitro* culture of the larval stages. *Research in Veterinary Science* **12**, 401–2.

RICKARD, M. D., COMAN, B. J. & CANNON, R. M. (1977). Age resistance and acquired immunity to *Taenia pisiformis* in dogs. *Veterinary Parasitology* **3**, 1–9.

ROTHWELL, T. L. W. (1978). Vaccination against the nematode *Trichostrongylus colubriformis*. III. Some observations on factors influencing immunity to infection in vaccinated guinea pigs. *International Journal for Parasitology* **8**, 33–7.

ROTHWELL, T. L. W. & LOVE, R. J. (1974). Vaccination against the nematode *Trichostrongylus colubriformis*. I. Vaccination of guinea pigs with worm homogenates and soluble products released during *in vitro* maintenance. *International Journal for Parasitology* **4**, 293–9.

SADUN, E. H. & LIN, S. S. (1959). Studies on the host parasite relationship to *Schistosoma japonicum*. IV. Resistance acquired by infection, by vaccination and by the injection of immune sera, in monkeys, rabbits and mice. *Journal of Parasitology* **45**, 543–8.

SINSKI, E. & HOLMES, P. H. (1977). *Nippostrongylus brasiliensis*: Systemic and local IgA and IgG immunglobulin responses in parasitized rats. *Experimental Parasitology* **43**, 382–9.

SMITH, M. A., CLEGG, J. A. & WEBBE, G. (1976). Cross-immunity to *Schistosoma mansoni* and *S. haematobium* in the hamster. *Parasitology* **73**, 53–64.

SMITH, W. D. (1977). Anti-larval antibodies in the serum and abomasal mucous of sheep hyperinfected with *Haemonchus contortus*. *Research in Veterinary Science* **22**, 334–8.

STROMBERG, B. E. (1979). The isolation and partial characterization of a protective antigen from developing larvae of *Ascaris suum*. *International Journal for Parasitology* **9**, 307–11.

STROMBERG, B. E., KHOURY, P. B. & SOULSBY, E. J. L. (1977). Development of larvae of *Ascaris suum* from the third to the fourth stage in a chemically defined medium. *International Journal for Parasitology* **7**, 149–51.

STROMBERG, B. E. & SOULSBY, E. J. L. (1977*a*). *Ascaris suum*: immunization with soluble antigens in the guinea pig. *International Journal for Parasitology* **7**, 287–91.

STROMBERG, B. E. & SOULSBY, E. J. L. (1977*b*). Heterologous helminth induced resistance to *Ascaris suum* in guinea pigs. *Veterinary Parasitology* **3**, 169–75.

TAYLOR, A. E. R. & MULLER, R. (1980). *Vaccines against Parasites*. London: Blackwell Scientific Publications.

TAYLOR, M. G. (1980). Vaccines against trematodes. In *Vaccines against Parasites*, (ed. A. E. R. Taylor and R. Muller), pp. 115–40. London: Blackwell Scientific Publications.

TAYLOR, M. G., JAMES, E. R., BICKLE, Q. D., HUSSEIN, M. F., ANDREWS, J. J., BOBINSON, A. R. & NELSON, G. S. (1979). Immunization of sheep against *Schistosoma bovis* using an irradiated schistosomular vaccine. *Journal of Helminthology* **53**, 1–5.

TAYLOR, M. G., NELSON, G. S., SMITH, M. & ANDREWS, B. J. (1979). Studies on heterologous immunity in schistosomiasis. VII. Observations on the development of acquired homologous and heterologous immunity to *Schistosoma mansoni* in baboons. *Bulletin of the World Health Organization* **49**, 59–65.

THOMAS, H. C. & PARROT, M. V. (1974). The induction of tolerance to a soluble protein antigen by oral administration. *Immunology* **27**, 631–9.

THOMPSON, R. C. A. (1976). Inhibitory effect of BCG on development of secondary hydatid cysts of *Echinococcus granulosus*. *Veterinary Record* **99**, 273.

TOMASI, T. B. (1976). *The Immune System of Secretions*. New Jersey: Prentice-Hall Foundations of Immunology Series.

TURNER, E. L., BERBERIAN, D. A. & DENNIS, E. W. (1936). The production of artificial immunity in dogs against *Echinococcus granulosus*. *Journal of Parasitology* **22**, 14–28.

Urquhart, G. M., McIntyre, W. I. M., Mulligan, W., Jarrett, W. F. G. & Sharpe, N. C. C. (1963). Vaccination against helminth disease. *Proceedings of the 17th International Veterinary Congress* **1**, 769–74. Hannover: Hahn-Druckerei.

Urquhart, G. M., Jarrett, W. F. H., Jennings, F. W., McIntyre, W. I. M. & Mulligan, W. (1966). Immunity to *Haemonchus contortus* infection: relationship between age and successful vaccination with irradiated larvae. *American Journal of Veterinary Research* **27**, 1645–8.

Valera-Diaz, V. (1970). The immune response to gastro-intestinal parasites in sheep. Ph.D. thesis, University of Pennsylvania.

Valera-Diaz, V. M., Williams, J. F., Coltorti, E. A. & Williams, C. S. F. (1974). Survival of cysts of *Echinococcus granulosus* after transplant into homologous and heterologous hosts. *Journal of Parasitology* **60**, 608–12.

Vernes, A. (1976). Immunization of the mouse and minipig against *Trichinella spiralis*. In *Biochemistry of Parasites and Host–Parasite Relationships*, (ed. H. van den Bossche), pp. 319–324. Amsterdam: North Holland.

Wakelin, D. (1978). Immunity to intestinal parasites. *Nature, London* **273**, 617–20.

Wickerhauser, T. (1974). Experience of vaccinating calves against bovine cysticercosis due to *Cysticercus bovis*. *Cahiers de Medicine Veterinaire* **43**, 95–9.

Wickerhauser, T., Zukovic, M. & Dzakula, N. (1971). *Taenia saginata* and *T. hydatigena*: intramuscular vaccination of calves with oncospheres. *Experimental Parasitology* **30**, 36–40.

Parasitology (1981), **83**, 435–449
With 3 figures in the text

Hatching mechanisms of nematodes

R. N. PERRY *and* A. J. CLARKE
Nematology Department, Rothamsted Experimental Station, Harpenden, Herts. AL5 2JQ

(*Accepted* 6 *January* 1981)

INTRODUCTION

Investigations of nematode hatching have concentrated on the parasitic forms, especially on synchronization between host and parasite life-cycles which, in some species, is so close that the parasite is dependent on the host for the hatching stimulus. Research into nematode hatching reflects the artificial separations created throughout nematology by defining research groups and their activities in terms of the host; consequently, useful comparisons between hatching of plant and animal parasitic nematodes have been few. With the increasing use by geneticists of nematodes as model animals and the interest in them by neurophysiologists, molecular biologists and others, we hope that such distinctions will be lessened.

Recently, advances have been made in understanding the underlying mechanisms of hatching and this article examines the main events, especially the responses of the juvenile, between the onset of the hatching process and the subsequent emergence of the juvenile from the egg; it does not examine the various factors, such as temperature and root diffusate, involved in the initiation of hatching. We aim to demonstrate the differences and establish the similarities in various events in the hatching sequence of nematodes and to indicate areas where future research could be useful.

IMPORTANCE OF THE EGG-SHELL

Information on the structure and function of the nematode egg-shell has been reviewed (Wharton, 1980) in the context of the survival advantages the shell confers on the unhatched nematode. Some of the attributes of the nematode egg concerned with survival of the enclosed juvenile prevent a simple hatching sequence, so that various alterations in the egg-shell and its contents must occur before the juvenile can escape. Such changes may involve the permeability of the egg-shell and it is possible that enzymes may be involved. The lipid layer probably determines the permeability characteristics of the egg-shell, so any changes in this layer are important.

(*i*) *Egg-shell permeability*

As early as 1911 Looss observed that eggs of *Ancylostoma duodenale* in strong salt solutions collapsed more easily when ready to hatch. Subsequently, research on other nematodes has often indicated that an initial stage of the hatching sequence is a change in permeability of the egg-shell. The nature of the permeability change is important: water and/or solute movement through the egg-shell may be involved, although the inference from much of the earlier work was that an egg-shell, previously impermeable to water, became permeable before hatching. Wallace (1966) suggested that the lipid layer in eggs of *Meloidogyne javanica* is dissolved by enzyme action, the egg-shell then becoming permeable to water. Wilson (1958), by determining the number of plasmolysed eggs in hypotonic solutions at various times after the start of the hatching sequence, demonstrated that eggs of *Trichostrongylus retortaeformis* became permeable to water shortly before the eggs hatched. During the pre-hatch development the eggs were completely impermeable to water. He considered that ions in the solution surrounding the egg controlled the rate of permeability change. Using the same method, Anya (1966) postulated that hatching of *Aspiculuris tetraptera* was also associated with an increase in the permeability of the egg-shell to water. The egg-shell of the rat pinworm, *Syphacia muris*, also becomes permeable to water during the initial stages of hatching and van der Gulden & van Aspert-van Erp (1976) considered this a necessary preliminary to triggering hatch. Several reports on other species provide indirect evidence (for example size increase of the egg) of changes in permeability during the hatching process. Direct evidence was presented by Fairbairn (1961) in his study of *Ascaris lumbricoides*.

The change in permeability of the ascaroside membrane (composed of the glycolipids (ascarosides) (Tarr & Fairbairn, 1973)) in *A. lumbricoides* is accompanied by a leakage of trehalose from the egg. Enzymes can then pass through the ascaroside membrane which remains intact, to attack the outer, largely protein and chitin layer of the egg-shell. Leakage of trehalose has been postulated as part of the hatching sequence of *Globodera rostochiensis* (Clarke, Perry & Hennessy, 1978; Perry, 1978) and the importance of this, in terms of the removal of osmotic stress on the juveniles, is discussed below (see *Water uptake* section). Although the induction of permeability in the ascaroside membrane does not cause obvious structural breakdown, the change is sufficient to allow the passage of large enzyme molecules. Barrett (1976), however, could detect no chemical changes in the ascaroside layer and suggested that the permeability change must be due either to very localized chemical or conformational changes or to mechanical damage by the activated juvenile.

The permeability properties of the *Ascaris* egg-shell require redefinition. Before hatching the egg-shell is reported to be permeable only to organic solvents and lipid-soluble and respiratory gases. However, work by Clarke & Perry (1980) showed that the water content of unhatched juveniles, which had not been stimulated to hatch, changed with the osmotic pressure of the medium in which the eggs were immersed, and that the effect was reversible. Thus, the egg-shell is permeable to water and acts as a semi-permeable membrane. It is impermeable to water-soluble molecules such as trehalose until the hatching sequence is initiated when the egg-shell alters to allow trehalose to leak into the external medium. In view of this, the nature of the egg-shell permeability change in other species should be investigated.

Permeability change (discussed in connection with cyst nematode hatching by Clarke & Perry (1977)) is one of several processes known to be calcium-mediated in other animals and the current interest in these cellular control processes gives new significance to the work of Ellenby & Gilbert (1957, 1958) on the cardiotonic activity of the hatching factor and the synergisitic effect of Ca^{2+} on the hatch of *G. rostochiensis*. Clarke, Cox & Shepherd (1967) showed that Ca^{2+} was a major inorganic constituent of isolated *G. rostochiensis* egg-shells. Atkinson & Ballantyne (1979) and Atkinson, Taylor & Ballantyne (1980) consider that Ca^{2+} has an active rôle in the hatching of this species. Low concentrations of lanthanum chloride and ruthenium red, considered to be specific inhibitors of calcium-mediated processes, inhibit hatching and ionophores which sequester Ca^{2+} and pass through membranes avoiding control systems, have a synergistic effect on sub-optimal concentrations of the hatching factor (Atkinson & Ballantyne, 1979). Both ruthenium red and lanthanum chloride are reported to bind to eggs treated with potato root diffusate (Atkinson & Taylor, 1980). Uptake of Ca^{2+} by the unhatched juvenile and the egg-shell when eggs are exposed to the hatching stimulus has been detected by Atkinson *et al.* (1980). These authors point out that uptake by the egg-shell may be an incidental occurrence as Ca^{2+} passes through to the enclosed juvenile, or it may be a direct effect by the hatching factor on the egg-shell. Recent evidence (Taylor & Atkinson, 1980) indicates that a major binding site of Ca^{2+} is located on the egg-shell.

There is evidence against the theory that hatching of *G. rostochiensis* involves Ca^{2+} transport. The postulated role of lanthanum chloride as an inhibitor seems vitiated by its hatching activity. Clarke & Hennessy (1981) tested a range of concentrations of ruthenium red and lanthanum chloride including those used by Atkinson & Ballantyne (1979). They found that 0·01 mM to 10 mM lanthanum chloride *initiated* the hatching of *G. rostochiensis*; 4 mM lanthanum chloride in distilled water, for example, had about the same hatching activity as the same concentrations of lanthanum chloride in root diffusate: both elicit nearly 40% hatch. Clarke & Hennessy (1981) confirmed Atkinson & Ballantyne's (1979) observations that ruthenium red inhibits the hatching of *G. rostochiensis*; however, the hatching activity of root diffusate was destroyed when ruthenium red was added to it and then removed and, therefore, failure to cause hatching may be due to inactivation of the hatching factor rather than interference with the hatching mechanism. Ruthenium red also inhibits nematode movement. Metal chelating agents may be used as inhibitors of cellular Ca^{2+} transport systems but Clarke & Hennessy (1981) obtained substantial hatching of *G. rostochiensis* in root diffusate from which Ca^{2+} had been removed and which contained up to 12 mM of the calcium-chelating agent 1,2-di(2-aminoethoxyl)ethane-*N*,*N*,*N*′,*N*′-tetra-acetic acid. Further investigations are required to establish the presence of a Ca^{2+} transport system associated with the hatching of this species.

(ii) *Enzymes*

Direct action of the root diffusate could possibly cause hatching of *G. rostochiensis*, with the hatching factor interfering with the close packing of the molecules of the lipid layer. However, if hatching factors can act at the extreme dilutions (10^{-14} g/ml) reported by Masamune (1976), it would suggest that they do not operate directly but by a response involving enzyme action. The involvement of enzymes in hatching has been postulated for a number of species. About 24 h before the hatching of juveniles of *Xiphenema diversicaudatum* the oesophageal bulb frequently pumps, and Flegg (1968) observed fluid leaving the mouth; the rigid egg membrane then apparently becomes flexible and the egg diameter increases. Taylor (1962) observed a rapid pulsation of the valve in the metacorpus of *Aphelenchus avenae* before hatching but could detect no fluid leaving the mouth or stylet; but it was only after these pulsations that the egg membrane became more flexible. Softening of the egg-shell during the hatching process has been described in many plant-parasitic nematodes (e.g. Wallace, 1968; Bridge, 1974) and usually the action of enzymes, possibly released by the nematode, is inferred. Research is required to obtain direct evidence of the degree of enzyme involvement and, if they are present, their origin and mode of activation. Enzymes need not necessarily be emitted by the juvenile but could already be present in the egg fluid or egg-shell, awaiting activation by the hatching factor or the removal of inhibitors by diffusion. In the absence of direct evidence of enzyme activity alternative explanations remain as possibilities: Thistlethwayte (1969) suggested that in *Pratylenchus penetrans* shell distortion occurred because the juvenile showed more vigorous movement and not because the egg-shell became more flexible. He concluded that enzymes played 'a small part, at best,' in the hatching of this species and that mechanical forces were mainly involved. Bird (1971) considered that enzymic breakdown of the lipid layer would allow the juvenile more freedom of movement; this may be associated with other events, such as water uptake by the juvenile (see *Water uptake* section).

Wilson (1958) suggested that the inner lipid layer of the egg-shell of *T. retortaeformis* was emulsified by the active movement of the juvenile, with assistance of emulsifying agents in the egg-fluid. Wallace (1966, 1968) considered that the lipid layer of *M. javanica* egg-shells could be altered either by emulsification or by the secretion of an enzyme which dissolves the lipid layer at about the time of the first juvenile moult. Ultrastructural studies of *Meloidogyne* (Bird, 1968) indicate that enzymes cause hydrolysis of the lipid layer although this does not occur at the time of the first juvenile moult but later when hatching commences. The enzymes appear to be synthesized in the sub-ventral oesophageal glands of the juvenile just before hatching. Bird (1968) suggested that the hemizonid of *M. javanica* may function as a receptor for hatching stimuli and trigger enzyme synthesis.

Rogers (1958, 1960) and Fairbairn (1961) demonstrated that hatching of *A. lumbricoides* involves enzymes. Chitinase and lipase activity were detected in the hatching medium after release of juveniles. Rogers (1958) suggested that a protease might also be involved. The enzyme was subsequently detected by Hinck & Ivey (1976) who showed that protease activity increased simultaneously with hatching. Indirect evidence was taken to indicate that the enzymes are released by the 2nd-stage juvenile (Ward & Fairbairn, 1972) but so far no secretory activities have been reported for this much studied species. It is possible that the enzymes are present in the egg-fluid and are kept inactive either by separation from their substrates by the lipid membrane or by an inhibitor. It would be interesting to know if the enzymes' activity is suppressed by trehalose at the concentrations (Fairbairn & Passey, 1957; Clarke & Perry, 1980) occurring in the egg-fluid; if so, trehalose release would precede the activity of enzymes responsible for eroding the egg-shell to facilitate eclosion. Permeability changes of the lipid layer would be caused by other enzymes, perhaps located elsewhere than the egg-fluid.

Rogers (1978) supports the view that *Ascaris* juveniles release the enzymes in direct response to the hatching stimulus. He outlined a mechanism for nematode hatching analogous to the process of juvenile exsheathment. This hypothesis suggests that the hatching stimulus (CO_2) affects sulphydryl groups in a receptor in the juvenile, leading via non-adrenergic activity to the release of a neurohormone which in turn causes the release of the enzymes which attack the egg-shell. The analogies between the hatching processes and exsheathment led Rogers (1978) to find that insect juvenile hormone analogues inhibit the hatching of *Nematospiroides dubius*, *Nippostrongylus brasiliensis*, *Haemonchus contortus*, *Nematodirus spathiger* and *Aphelenchus avenae* and he suggested that the compounds may affect either the release or the action of the putative neurohormone.

Structural specialization of the egg-shell at one or both poles to form an operculum occurs in some nematode species. The chitinous seal between the operculum and the egg-shell of *Syphacia muris* is dissolved after change in egg-shell permeability (van der Gulden & van Aspert-van Erp, 1976) allowing the operculum to open. The operculum of *Aspiculuris tetraptera* consists of a modification of the uterine and chitinous layer of the egg-shell over the whole area of the operculum (Wharton, 1979*a*) while in *S. obvelata* the modification occurs at the groove which delimits the operculum (Wharton, 1979*b*). Eggs of *Trichuris suis* have an opercular plug of a chitin/protein complex at each pole, different in organization from the rest of the shell (Wharton & Jenkins, 1978). Both plugs are dissolved before hatching and the juvenile uses its stylet to pierce the lipid layer and emerge head first (Beer, 1973). Wharton & Jenkins (1978) suggested that, because chitin which is bound to protein is not degradable by chitinase, the lower proportion of protein combined with the chitin/protein arrangement of the opercular plugs makes them more susceptible to enzymic attack. It is surprising that more studies have not been carried out on the initiation, nature and sequence of the enzymic reaction with the operculum and its relation to the hatching process as a whole.

CONSEQUENCES OF PERMEABILITY CHANGE

(i) *Differences in hatching patterns*

Hatching of *Ascaris* eggs may be close to 100 % during a single exposure to the hatching stimulus; there is no indication of limited hatching (cf. cyst nematodes) followed by further hatching on re-stimulation. The pattern accords with the parasite's life-cycle where no advantage accrues by delaying eclosion once the eggs have been ingested. Where a single exposure (of several hours) to the hatching stimulus can cause nearly complete hatching it suggests that the juvenile itself need not necessarily be directly implicated in the early stages of the process; possibly only changes in the egg-shell permeability are initially involved. Possible causes of the permeability change are discussed above. Barrett (1976) also suggested that the change might be initiated by mechanical damage of the membrane caused by the movement of the juvenile. Clarke & Perry (1980), however, concluded that enhanced juvenile activity occurs as a *result* of membrane permeability change with the consequent release of trehalose and removal of osmotic stress. It is not clear whether the unhatched *Ascaris* juvenile is involved in the initial stages of the hatching sequence before the change in egg-shell permeability.

In contrast, however, it can be argued that the hatching pattern of *G. rostochiensis* indicates that the juvenile of this species is involved at an early stage in the sequence. If the hatching factor (in root diffusate) alters the permeability of the egg-shell (either directly or indirectly but without involvement of the juvenile) then all viable eggs would hatch if immersed long enough in root diffusate. This does not accord with the hatching pattern of this organism; some eggs are left unhatched many of which will hatch only if re-stimulated. Eggs of the related species, *G. pallida*, when exposed to root diffusate for just 5 min per week for 4 weeks gave over 40 % hatch, whereas continuous exposure to diffusate, replenished each week, increased hatch by less than 20 %; eggs in water, without exposure to diffusate, gave less than 10 % hatch (Forrest & Perry, 1980). Thus, the diffusate acts very rapidly to initiate hatching; duration of stimulation has little importance. Experiments with *G. rostochiensis* (Perry & Beane, unpublished results) substantiate and extend this work: short exposures to diffusate (minimum 5 min) each week for 5 weeks elicits the same level of hatch (60–70 %) as continuous exposure to diffusate, replenished each week, and there is evidence that it is the *repetition* of the stimulation which is of major importance for obtaining close to 100 % hatch. Accordingly, it is more logical to consider the hatching sequence of this species as involving an interaction between the stimulus and the juvenile at an early phase in the sequence. Thus, juveniles not in a physiological state to hatch–in diapause, for example–would still be retained in an unaltered egg-shell. Involvement of the juvenile need not necessarily imply active movement, especially if enzyme secretions occur. Enhanced movement would be possible only after the permeability change has allowed increased hydration. Ellenby & Perry (1976) suggested that the *G. rostochiensis* juvenile could be involved in the initial stages of the hatching sequence, as the hatching factor may initiate a neurosecretory response from the juvenile.

In contrast to the behaviour of *G. rostochiensis*, Croll (1974) observed that in *Necator americanus* head waving is one of the first actions in the hatching process; movement of the anterior end and of the whole body increases about the same time or before an increase in egg volume occurs. Croll (1974) considered 'feeding' movements to be instrumental in flushing enzymes into the egg via the anus, although this was not observed. He suggested that an influx of water occurred as a result of the movement of the juvenile and the increase in pressure thus generated within the egg was instrumental in bursting it and letting the juvenile escape.

The contrasts between species outlined above, highlight at least two main differences in the hatching process which may obviate a 'model' hatching system universal to all nematodes. The first difference is in the nature of the impermeability of the egg-shell and whether the egg-shell is flexible or remains rigid during the later stages of eclosion; the second is in the rôle of water uptake in hatching.

(ii) *Water uptake*

Several workers (Wilson, 1958; Croll, 1974; van der Gulden & van Aspert-van Erp, 1976) have inferred from indirect evidence that there is water movement into the egg and/or juvenile at some stage before eclosion. Direct evidence for increased hydration of the nematode before hatching was first obtained by Ellenby & Perry (1976) using *G. rostochiensis*. When wet, the egg-shell of this species is permeable to water in both directions (Ellenby, 1968) yet Ellenby & Perry (1976) found that the unhatched juvenile is maintained in a state of incomplete hydration with a water content of 67% and that the water content of the juvenile increases significantly before hatching when eggs are exposed to the hatching stimulus, potato-root diffusate. The significance of this observation is increased by the finding of Clarke & Hennessy (1976) that the egg-fluid contains trehalose at a concentration of about 0·34 M; the concentration of trehalose is thought to be important in the regulation of juvenile water content and, perhaps, the maintenance of dormancy. Using interference microscopy, Clarke *et al.* (1978) showed that the water content of active, hatched juveniles decreased when they were transferred from distilled water to trehalose or sucrose solutions of different concentrations, the water loss increasing with solute concentration (Fig. 1). The water contents stabilize after 24 h in the solutions, and in the 0·4 M concentration (close to that of the egg-fluid) the juvenile water content is 67%, the same as that determined for unhatched, unstimulated individuals in eggs equilibrated with water. Other experiments (Clarke *et al.* 1978) demonstrated that the incomplete hydration of the unhatched juvenile limits movement and suppresses hatch. Clarke & Perry (1977) proposed that a change in permeability of the egg-shell allowed the diffusion of egg-fluid solutes out of the egg, which removed the osmotic stress on the juvenile with concomitant increase in juvenile water content. This increase is needed for *G. rostochiensis* juveniles to become fully mobile and start the behavioural sequence (see below) leading to hatch from the egg.

An interesting comparison can be made with *Heterodera schachtii*. The unhatched, unstimulated juvenile of this species contains more water than that of a comparable *G. rostochiensis* juvenile and there is no evidence of any water uptake prior to hatching (Perry, 1977*a*); the juvenile appears to be in a condition suitable for hatching without the application of root diffusate. Perry, Clarke & Hennessy (1980) showed that the egg fluid may not provide as great an osmotic stress as that found in *G. rostochiensis* eggs (Fig. 1) and that *H. schachtii* juveniles remain motile in solution concentrations that inhibit movement of *G. rostochiensis* juveniles. These two attributes may explain why *H. schachtii* hatches more readily in water than some other species. However, there is an additional hatch induced by root diffusates and perhaps a slight dilution of the egg fluid may be important in either decreasing the low osmotic pressure or removing possible inhibitors in the egg fluid which may be preventing hatch of a minority of juveniles. If the first explanation is correct it indicates that the ensuing water uptake was so slight as to be undetectable by interference microscopy. A similar situation could exist in *Nematodirus battus* where trehalose has been detected in the egg fluid (Ash & Atkinson, 1980) but there is no water uptake prior to hatch (Perry, 1977*b*); it would be interesting to know the concentration of trehalose present. Alternatively, the nematodes could be adapted to the osmotic pressure.

These ideas on hatching of *G. rostochiensis* assume a change in the permeability of the egg-shell with associated leakage of trehalose, caused directly or indirectly by hatching agents; this has yet to be established. Juvenile hatching occurs over weeks, which presents problems in detecting trehalose leakage in a non-sterile medium. However, these stages in the hatching process have been demonstrated

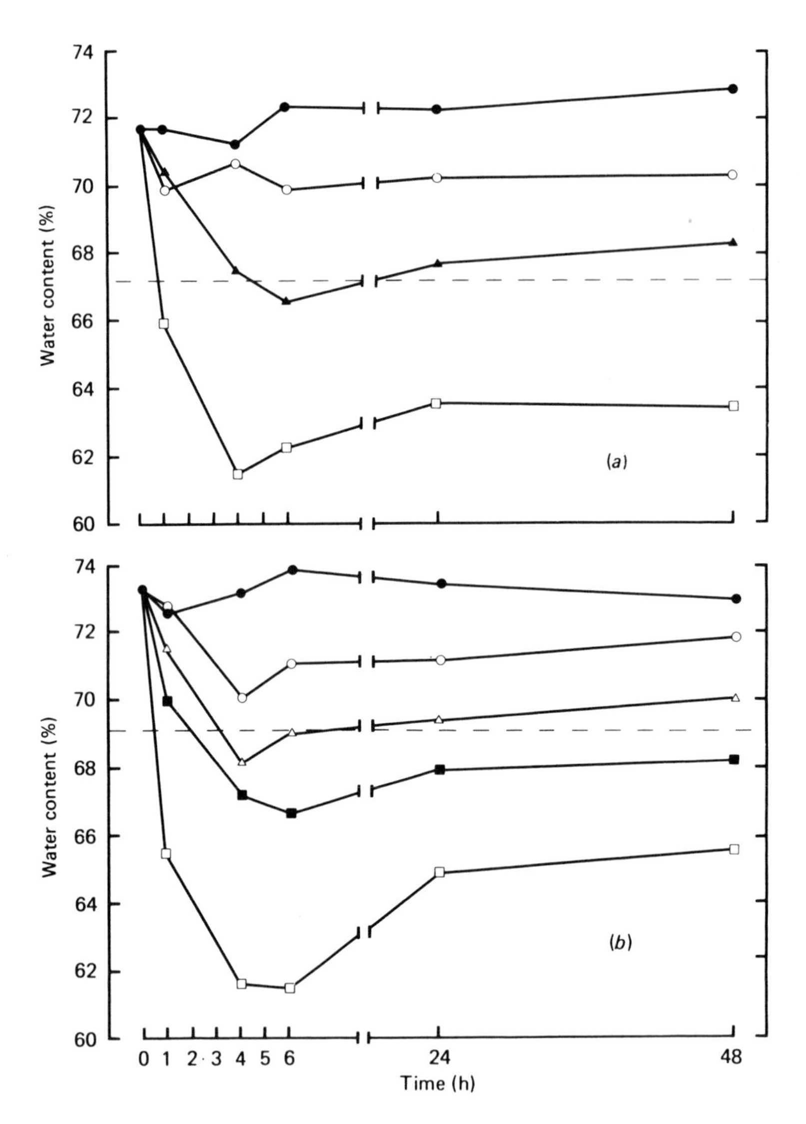

Fig. 1. Changes in water content of 2nd-stage juveniles of *Globodera rostochiensis* (*a*) and *Heterodera schachtii* (*b*) in glass distilled water (GDW) and after transfer from distilled water to various trehalose solutions. The broken line is the water content of unhatched, unstimulated juveniles of each species. Trehalose concentrations: (○), 0·2 M; (△), 0·3 M; (▲), 0·38 M; (■), 0·4 M; (□), 0·6 M and (●), GDW (from Clarke, Perry & Hennessy, 1978 and Perry, Clarke & Hennessy, 1980).

in *Ascaris* and their importance has now been examined. Clarke & Perry (1980) suggested that the egg-shell of *A. suum* changed from a semi-permeable to a permeable membrane after application of the hatching stimulus. The unhatched juvenile is then gradually exposed to a new environment as the egg fluid solutes diffuse outwards and there is an associated uptake of water by the juvenile before hatching (Fig. 2). The osmotic pressure of the egg fluid is equivalent to between 0·1 and 0·2 M trehalose (Clarke & Perry, 1980), which is less than the 0·4 M trehalose concentration calculated by Fairbairn & Passey (1957). The water content of the

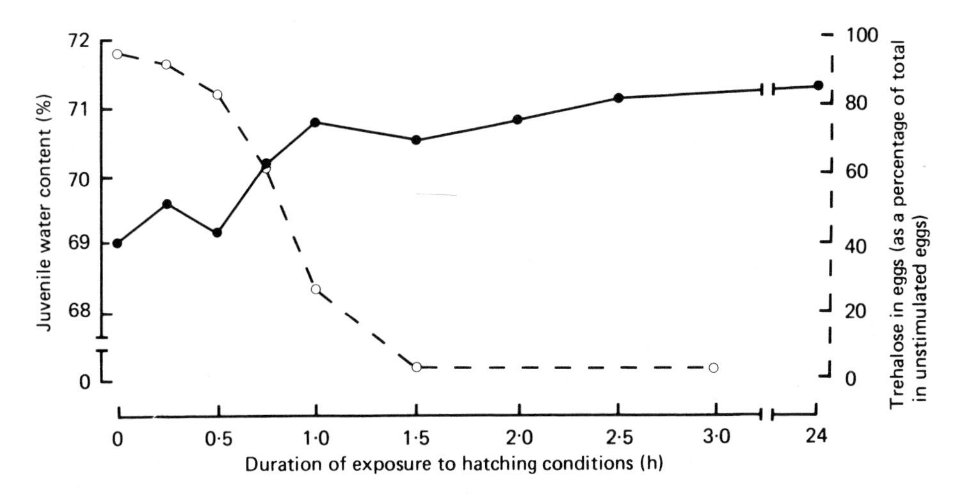

Fig. 2. Changes in the water content of unhatched juveniles (●) and in the trehalose content of eggs (○) of *Ascaris* at 38 °C during the hatching process. Water content of unhatched juveniles was determined at set time intervals after the removal of eggs from Fairbairn's hatching medium continuously gassed with $N_2:CO_2$ (95:5 v/v) (from Clarke & Perry, 1980). Trehalose leakage was determined by Fairbairn (1961) after exposure of eggs to hatching conditions for set times; he expressed this as a percentage of the total carbohydrate extractable from unstimulated eggs. We have subtracted these values from 100 to estimate the trehalose content of the eggs.

unhatched, unstimulated juveniles of *A. suum* is the same as that of *H. schachtii* juveniles but greater than that of *G. rostochiensis* juveniles. The increase in hydration of *A. suum* may be responsible for ending quiescence of the juvenile and it is likely, as with some other nematodes before hatching, that an increase in volume and turgor pressure would aid the juvenile in breaking out of the egg-shell.

(iii) *Other physiological changes*

When the dormant *Ascaris* egg is stimulated the oxygen quotient increases (Passey & Fairbairn, 1955) and the relatively large ATP/ADP ratio falls (Beis & Barrett, 1975). Similar changes were reported for *G. rostochiensis* eggs (Atkinson & Ballantyne, 1977*a*, *b*); they occur during the first 24 h after stimulation while the nematode water content is increasing (Ellenby & Perry, 1976). Beis & Barrett (1975) found no evidence of any general metabolic inhibitor in dormant *Ascaris* eggs, a result consistent with dormancy being maintained by a water deficit (Clarke & Perry, 1980) induced by the egg-fluid solutes. There is no evidence that a major alteration of metabolic pathway occurs in nematodes as a result of the change from the dormant to the active, hatching state.

(iv) *Behaviour leading to eclosion*

Cinématographic studies by Doncaster & Shepherd (1967) and the careful analysis of behavioural sequences (Doncaster & Seymour, 1973) have provided a clear demonstration of the activity phases of *G. rostochiensis* juveniles before

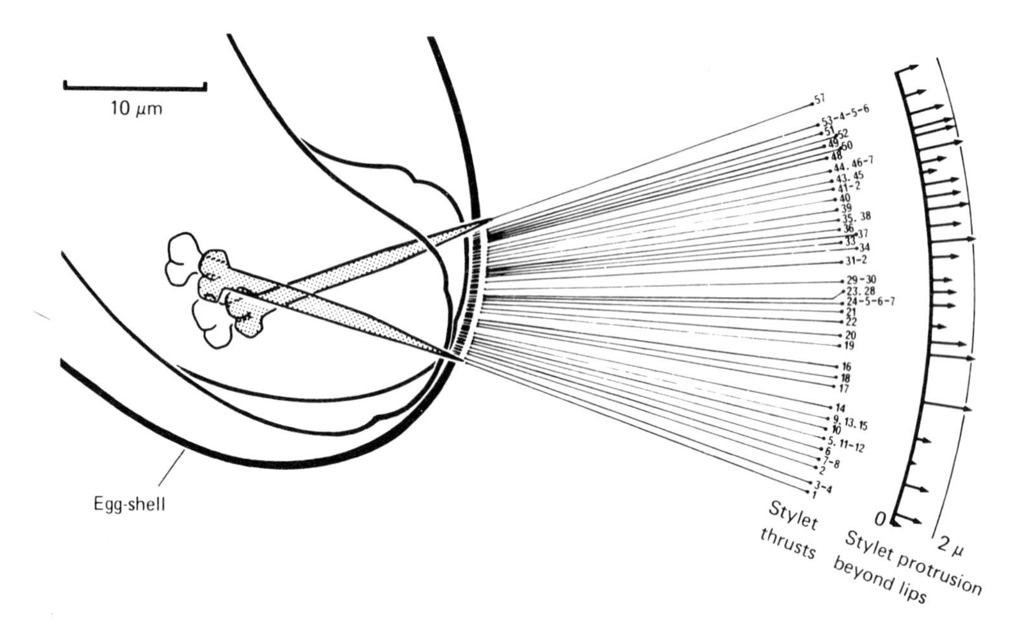

Fig. 3. Diagram, based on analysis of ciné film, of a *Globodera rostochiensis* 2nd-stage juvenile using its stylet to perforate the egg-shell and make part of the exit slit. Outlines of head and stylet (protracted and retracted) at the start and end of the sequence are shown. Within 80 sec 57 stylet thrusts were made; in 26 of these the stylet perforated the shell. Each arrow shows where a perforation occurred and how far the stylet then protruded beyond the lips. As the numbering of the thrusts shows, when the stylet failed to penetrate it was thrust again (up to 4 times) nearer the previous hole until it penetrated, when it usually broke into a previous perforation. After this sequence the rest of the slit was made in the opposite direction, beginning near its original starting point. (From Doncaster & Seymour, 1973.)

eclosion. The first of three phases, termed 'widespread exploration' by Doncaster & Seymour (1973), involves the juvenile circulating within the egg; this is followed by 'local exploration' when only the head end moves and the lip region is pressed and/or rubbed against the egg-shell which is also probed by the stylet. The third phase, 'stylet thrusting', involves co-ordinated, accurate stylet thrusts causing perforation of the egg-shell and subsequent formation of a slit through which the juvenile escapes (Fig. 3); Doncaster & Seymour (1973) termed the alternation of the last two phases the 'cutting cycle'. The use of the word 'cutting' is somewhat misleading: the slit in the egg-shell is not caused by an edged instrument but the tip of the stylet makes a line of perforations, each hole being close enough together for the stylet conus to break through to the preceding one (Doncaster & Shepherd, 1967). The stylet thrusts were slow; one juvenile made 43 thrusts/min and the slit in the egg-shell enlarged at 7·5 μm/min (Doncaster & Seymour, 1973). By pressing against the side of the egg-shell, the nematode causes the slit to gape and probably one or two of the labial tactors then lose contact with the egg-shell and perception of a different stimulus may determine the direction of head movement (Doncaster, 1974). There is no indication that the juvenile moves rapidly through the slit under the influence of increased pressure within the egg, indeed any increase in internal pressure of the egg fluid is likely to have been lost, perhaps by the stylet thrusts penetrating the egg-shell (unless the holes are self-sealing) but certainly when the holes run together to form the slit. Taylor (1962) noted that when the stylet of *Aphelenchus avenae* punctured the egg-shell, the egg almost immediately became less turgid and the end opposite the punctures collapsed except when supported by the juvenile.

The formation of a slit from a line of perforations may be necessary in *G. rostochiensis* because the egg-shell remains rigid throughout the hatching process. Other plant-parasitic nematodes use the stylet differently in hatching behaviour. *P. penetrans* did not make a series of holes but one stylet thrust of many was able to penetrate the stretched shell and this was apparently sufficient to enable the nematode's head to initiate a tear in the shell (Thistlethwayte, 1969). This is similar to *H. avenae* where stylet thrusting appeared to weaken or penetrate the egg-shell enabling it to be split by the force of the juvenile head moving slowly forward (Banyer & Fisher, 1972). In *Tylenchorhynchus maximus* and *Merlinius icarus* there is an increase in size of eggs about 12 h prior to hatch but juveniles are not ejected from eggs under pressure; escape appears to be due to mechanical forces as the juvenile body presses against the egg-shell with the stylet only used to initiate the rupture of the stretched, flexible egg-shell (Bridge, 1974 and personal communication). This behaviour is more typical of plant-parasitic nematodes than the careful cutting process of *G. rostochiensis* and has been described, for example, in *M. arenaria* (Dropkin, Martin & Johnson, 1958), *M. javanica* (Bird, 1968) and *T. dubius* (Sharma, 1971). Wallace (1968) considered that the stimulus for stylet activity during hatching of *M. javanica* occurs when juvenile movement is impeded by an increase in friction between the nematode and the egg-shell caused by pressure increases within the egg. As juveniles of a related species, *M. incognita*, are fully hydrated when they leave the egg (Ellenby, 1974), the increase in *M. javanica* egg size noted by Wallace (1966) and the increase in pressure that he suggested, may be due to increased hydration of the juvenile after the permeability properties of the egg membrane have altered (Wallace, 1966). The juvenile head is pushed out making a protuberance in the flexible egg-shell and a series of stylet thrusts eventually causes a tear which is extended by the nematode as it escapes by bodily movement (Wallace, 1968).

A more detailed quantitative assessment of a similar hatching process has been carried out with *Ditylenchus dipsaci* (Doncaster & Seymour, 1979). The egg-shell is weakened in the region where the slit eventually appears by several hundred stylet thrusts at the rate of about 2/sec; they rarely penetrate the egg-shell, occur in bursts interspersed with anterior body movements and seem relatively haphazard compared with those of *G. rostochiensis*. However, as only 7 % of the stylet thrusts occur outside the position of the eventual slit, a high level of accuracy is involved; many thrusts are directed at the points which will form the ends of the slit. By pressing its head against the weakened region of the egg-shell, the juvenile causes the slit to open and the head emerges rapidly at about 54 μm/sec, declining to 10–20 μm/sec within 0·6 sec. The egg-shell does not collapse and occasionally an avoidance reflex may occur when the juvenile reverses back into the shell before finally leaving it, as was also noted by Wallace (1968) with *M. javanica*.

In some species the stylet does not appear to play any part in hatching. Juveniles of *Criconemoides xenoplax* become active just before hatching and the egg-shell suddenly breaks allowing the head to come out; the stylet is not involved in the process and it takes about 10–15 min for the juvenile to escape completely from the egg (Seshadri, 1964). No stylet activity was observed during the hatch of *Heterodera iri*; the juvenile ruptured the egg-shell with its tail tip and emerged tail-first (Laughlin, Lugthart & Vargas, 1974). Animal-parasitic nematodes rarely have well-developed organs for the mechanical rupture of the egg-shell and Silverman & Campbell (1959) considered that the tail of *Haemonchus contortus* is similarly used to make a hole in the shell through which the juvenile escapes tail-first. However, other reports suggest that the egg-shell bursts (Looss, 1911), or that the juvenile forces its way out by bodily movements (Veglia, 1915). Rogers & Brooks (1977) found that the chitin and/or protein of the egg-shell broke down in a localized region and, by physical force, the juveniles broke through the inner lipid membrane emerging head first.

In the above examples influx of water into the egg is frequently inferred from observations on the egg during hatching. However, in plant-parasitic nematodes especially, osmotically increased internal pressure of the egg fluid appears rarely to be involved with expulsion of juveniles and, as mentioned above, such pressure is likely to be lost once the egg is pierced. Although any increase in internal pressure may assist in causing a tear in the shell, the increase in egg size and, presumably, internal pressure may be caused by an increase in turgor and size of the juveniles as they take up water (see *Water uptake* section). Juveniles of *T. retortaeformis*, for example, become quiescent before hatching and Wilson (1958) considered that the egg-shell is burst by the increase in hydrostatic pressure of the nematode. In contrast, the build-up of internal pressure, apparently resulting from influx of water into the egg fluid, was considered by Croll (1974) to be instrumental in hatching of *N. americanus*. The nematode head is most frequently found pressed firmly against the egg-shell near the pole and the rotary movements of the stoma causes a break in the egg-shell through which the juvenile is rapidly expelled. When eggs were observed to burst at a point away from the nematode's head, the egg collapsed, trapping the juvenile and preventing emergence (Croll, 1974). Development of a high internal pressure within the egg was also considered by Vinayak, Dhir, Chitkara & Naik (1978) to be essential to the emergence of juveniles of *Ancyclostoma duodenale*.

Understanding of the hatching mechanisms, especially the earlier biochemical and physiological changes to the egg-shell and juvenile respectively, will be improved if juvenile activity before eclosion is examined. Many of the earlier reports were largely anecdotal. More studies involving quantitative analyses of behavioural sequences are required before differences between species in the method of escape from the egg-shell can be assessed with confidence. The importance of such studies would be enhanced if they were linked, for example, to direct investigations of the possible occurrence and rôle in eclosion of water movement into the egg-fluid and/or the unhatched nematode.

(v) *Events immediately after eclosion*

Ellenby (1974) showed that the water content of the *G. rostochiensis* juvenile increased immediately after hatching from the egg. Clearly, although the juvenile takes up water before hatching, the egg-shell physically restricts its water content and the juvenile can only become fully hydrated when it escapes from the egg. The juvenile of this species fits the egg-shell completely, in contrast to *M. incognita* juveniles which have about 30% free space inside the egg and are fully hydrated before they hatch. An increase in water content immediately after hatching has also been demonstrated in *H. schachtii* (Perry, 1977*a*) and *N. battus* (Perry, 1977*b*).

CONCLUSION

It is important to know more about the hatching mechanisms of economically important nematodes so that possibilities for preventing hatch or for inducing it when conditions are unfavourable for nematode survival can be fully evaluated. *G. rostochiensis* and *G. pallida*, for example, cause yield losses of potatoes of some 9% annually in the U.K. (a cost equivalent in 1975 of £33 million) and are among the most serious agricultural pests in temperate regions (Stone, 1977).

We hope this article has indicated gaps in our present knowledge, has shown that understanding the phenomenon of hatching will benefit if the studies on plant-and animal-parasitic nematodes can be more closely linked, and has demonstrated the differences, as well as the similarities, in the various events during the hatching of nematodes.

We thank Mr C. C. Doncaster and Dr M. K. Seymour for permission to reproduce Fig. 3 and Professor D. Fairbairn for permission to use data contained in Fig. 2.

REFERENCES

Anya, A. O. (1966). Experimental studies on the physiology of hatching of eggs of *Aspiculuris tetraptera* Schulz (Oxyuroidea; Nematoda). *Parasitology* **56**, 733–44.

Ash, C. P. J. & Atkinson, H. J. (1980). Physiological events associated with hatching and dormancy in *Nematodirus battus* (Trichostrongyloidea). *Parasitology* **81**, x.

Atkinson, H. J. & Ballantyne, A. J. (1977*a*). Changes in the oxygen consumption of cysts of *Globodera rostochiensis* associated with the hatching of juveniles. *Annals of Applied Biology* **87**, 159–66.

Atkinson, H. J. & Ballantyne, A. J. (1977*b*). Changes in the adenine nucleotide contents of cysts of *Globodera rostochiensis* associated with the hatching of juveniles. *Annals of Applied Biology* **87**, 167–74.

Atkinson, H. J. & Ballantyne, A. J. (1979). Evidence for the involvement of calcium in the hatching of *Globodera rostochiensis*. *Annals of Applied Biology* **93**, 191–8.

Atkinson, H. J. & Taylor, J. D. (1980). A calcium-binding site on the eggshell of *Globodera rostochiensis* with a role in hatching. (Abstract) *European Society Nematologists XVth Symposium, Bari, Italy*, pp. 19–20.

Atkinson, H. J., Taylor, J. D. & Ballantyne, A. J. (1980). The uptake of calcium prior to the hatching of the second-stage juvenile of *Globodera rostochiensis*. *Annals of Applied Biology* **94**, 103–9.

Banyer, R. J. & Fisher, J. M. (1972). Motility in relation to hatching of eggs of *Heterodera avenae*. *Nematologica* **18**, 18–24.

Barrett, J. (1976). Studies on the induction of permeability in *Ascaris lumbricoides* eggs. *Parasitology* **73**, 109–21.

Beer, R. J. S. (1973). Studies on the biology of the life-cycle of *Trichuris suis* Schrank, 1788. *Parasitology* **67**, 253–62.

BEIS, I. & BARRETT, J. (1975). Energy metabolism in developing *Ascaris lumbricoides* eggs. II. The steady state content of intermediary metabolites. *Developmental Biology* **42**, 188–95.

BIRD, A. F. (1968). Changes associated with parasitism in nematodes. III. Ultrastructure of the egg shell, larval cuticle, and contents of the subventral oesophageal glands in *Meloidogyne javanica*, with some observations on hatching. *Journal of Parasitology* **54**, 475–89.

BIRD, A. F. (1971). *The Structure of Nematodes*. New York and London: Academic Press.

BRIDGE, J. (1974). Hatching of *Tylenchorhynchus maximus* and *Merlinius icarus*. *Journal of Nematology* **6**, 101–2.

CLARKE, A. J., COX, P. M. & SHEPHERD, A. M. (1967). The chemical composition of the eggshells of the potato cyst-nematode, *Heterodera rostochiensis* Woll. *The Biochemical Journal* **104**, 1056–60.

CLARKE, A. J. & HENNESSY, J. (1976). The distribution of carbohydrates in cysts of *Heterodera rostochiensis*. *Nematologica* **22**, 190–5.

CLARKE, A. J. & HENNESSY, J. (1981). Calcium inhibitors and the hatching of *Globodera rostochiensis*. *Nematologica* **27** (in the Press).

CLARKE, A. J., & PERRY, R. N. (1977). Hatching of cyst-nematodes. *Nematologica* **23**, 350–68.

CLARKE, A. J., & PERRY, R. N. (1980). Egg-shell permeability and hatching of *Ascaris suum*. *Parasitology* **80**, 447–56.

CLARKE, A. J., PERRY, R. N. & HENNESSY, J. (1978). Osmotic stress and the hatching of *Globodera rostochiensis*. *Nematologica* **24**, 384–92.

CROLL, N. A. (1974). *Necator americanus*: activity patterns in the egg and the mechanism of hatching. *Experimental Parasitology* **35**, 80–5.

DONCASTER, C. C. (1974). *Heterodera rostochiensis* (Nematoda) egg-hatch. *Encyclopaedia cinematographia E2035. Institut für den wissenschaftlichen Film, Göttingen.*

DONCASTER, C. C. & SEYMOUR, M. K. (1973). Exploration and selection of penetration site by Tylenchida. *Nematologica* **19**, 137–45.

DONCASTER, C. C. & SEYMOUR, M. K. (1979). Hatching of the stem nematode. *Rothamsted Experimental Station, Report for 1978* **1**, 186.

DONCASTER, C. C. & SHEPHERD, A. M. (1967). The behaviour of second-stage *Heterodera rostochiensis* larvae leading to their emergence from the egg. *Nematologica* **13**, 476–8.

DROPKIN, V. H., MARTIN, G. C. & JOHNSON, R. W. (1958). Effect of osmotic concentration on hatching of some plant parasitic nematodes. *Nematologica* **3**, 115–26.

ELLENBY, C. (1968). Desiccation survival in the plant parasitic nematodes, *Heterodera rostochiensis* Wollenweber and *Ditylenchus dipsaci* (Kühn) Filipjev. *Proceedings of the Royal Society*, B **169**, 203–13.

ELLENBY, C. (1974). Water uptake and hatching in the potato cyst nematode, *Heterodera rostochiensis*, and the root knot nematode, *Meloidogyne incognita*. *Journal of Experimental Biology* **61**, 773–9.

ELLENBY, C. & GILBERT, A. B. (1957). Cardiotonic activity of the potato-root eelworm hatching factor. *Nature, London* **180**, 1105–6.

ELLENBY, C. & GILBERT, A. B. (1958). Influence of certain inorganic ions on the hatching of the potato root eelworm, *Heterodera rostochiensis* Wollenweber. *Nature, London* **182**, 925–6.

ELLENBY, C. & PERRY, R. N. (1976). The influence of the hatching factor on the water uptake of the second stage larva of the potato cyst nematode *Heterodera rostochiensis*. *Journal of Experimental Biology* **64**, 141–7.

FAIRBAIRN, D. (1961) The *in vitro* hatching of *Ascaris lumbricoides* eggs. *Canadian Journal of Zoology* **39**, 153–62.

FAIRBAIRN, D. & PASSEY, R. F. (1957). Occurrence and distributions of trehalose and glycogen in the eggs and tissues of *Ascaris lumbricoides*. *Experimental Parasitology* **6**, 566–74.

FLEGG, J. J. M. (1968). Embryogenic studies of some *Xiphinema* and *Longidorus* species. *Nematologica* **14**, 137–45.

FORREST, J. M. S. & PERRY, R. N. (1980). Hatching of *Globodera pallida* eggs after brief exposures to potato root diffusate. *Nematologica* **26**, 130–2.

GULDEN, W. J. I. VAN DER & ASPERT-VAN ERP, A. J. M. VAN. (1976). *Syphacia muris*: water permeability of eggs and its effect on hatching. *Experimental Parasitology* **39**, 40–4.

HINCK, L. W. & IVEY, M. H. (1976). Proteinase activity in *Ascaris suum* eggs, hatching fluid, and excretions-secretions. *Journal of Parasitology* **62**, 771–4.

LAUGHLIN, C. W., LUGTHART, J. A. & VARGAS, J. M. (1974). Observations on the emergence of *Heterodera iri* from the egg. *Journal of Nematology* **6**, 100–1.

LOOSS, A. (1911). The anatomy and life-history of *Ancylostoma duodenale* Dub. Part II. *Records of the Egyptian Government School of Medicine* **4**. (Monograph, 446 pp).

MASAMUNE, T. (1976). Purification of the hatching substance of the soybean cyst nematode. *Abstract of Special lecture at 34th Annual Meeting Chemical Society of Japan* p. 91.

PASSEY, R. F. & FAIRBAIRN, D. (1955). The respiration of *Ascaris lumbricoides* eggs. *Canadian Journal of Biochemistry and Physiology* **35**, 511–35.

PERRY, R. N. (1977*a*). Water content of the second-stage larva of *Heterodera schachtii* during the hatching process. *Nematologica* **23**, 431–7.

PERRY, R. N. (1977*b*). A reassessment of the variations in the water content of the larvae of *Nematodirus battus* during the hatching process. *Parasitology* **74**, 133–7.

PERRY, R. N. (1978). Events in the hatching process of the potato-cyst nematode *Globodera rostochiensis*. *Agricultural Research Council Research Review* **4**, 79–83.

PERRY, R. N., CLARKE, A. J. & HENNESSY, J. (1980). The influence of osmotic pressure on the hatching of *Heterodera schachtii*. *Revue de Nématologie* **3**, 3–9.

ROGERS, W. P. (1958). Physiology of the hatching of eggs of *Ascaris lumbricoides*. *Nature, London* **181**, 1410–11.

ROGERS, W. P. (1960). The physiology of infective processes of nematode parasites: the stimulus from the animal host. *Proceedings of the Royal Society*, B **152**, 367–86.

ROGERS, W. P. (1978). The inhibitory action of insect juvenile hormone on the hatching of nematode eggs. *Comparative Biochemistry and Physiology* A **61**, 187–90.

ROGERS, W. P. & BROOKS, F. (1977). The mechanism of hatching of eggs of *Haemonchus contortus*. *International Journal for Parasitology* **7**, 61–5.

SESHADRI, A. R. (1964). Investigations on the biology and life cycle of *Criconemoides xenoplax*, Raski, 1952 (Nematoda: Criconematidae). *Nematologica* **10**, 540–62.

SHARMA, R. D. (1971). Studies on the plant parasitic nematode *Tylenchorhynchus dubius*. *Mededelingen Landbouwhogeschool, Wageningen* **71**, 1–45.

SILVERMAN, P. H. & CAMPBELL, J. A. (1959). Embryonic and larval development of *Haemonchus contortus* at constant conditions. *Parasitology* **49**, 23–38.

STONE, A. R. (1977). Cyst-nematodes–most successful parasites. *New Scientist* **76**, 355–6.

TARR, G. E. & FAIRBAIRN, D. (1973). Ascarosides of the ovaries and eggs of *Ascaris lumbricoides* (Nematoda). *Lipids* **8**, 7–16.

TAYLOR, D. P. (1962). Effect of temperature on hatching of *Aphelenchus avenae* eggs. *Proceedings of the Helminthological Society of Washington* **29**, 52–4.

TAYLOR, J. D. & ATKINSON, H. J. (1980). A calcium-binding site on the egg-shell of *Globodera rostochiensis* with a role in hatching. *Parasitology* **81**, xvii.

THISTLETHWAYTE, B. (1969). Hatch of eggs and reproduction of *Pratylenchus penetrans* (Nematoda: Tylenchida). Ph.D. thesis, Cornell University.

VEGLIA, F. (1915). The anatomy and life-history of the *Haemonchus contortus*. *Third and Fourth Reports of the Director of Vetarinary Research, Pretoria, S. Africa*, pp. 349–500.

VINAYAK, V. K., DHIR, S., CHITKARA, N. L. & NAIK, S. R. (1978). Mechanisms of hatching of larvae of *Ancylostoma duodenale*. *Indian Journal of Parasitology* **2**, 7–9.

WALLACE, H. R. (1966). The influence of moisture stress on the development, hatch and survival of eggs of *Meloidogyne javanica*. *Nematologica* **12**, 57–69.

WALLACE, H. R. (1968). Undulating locomotion of the plant parasitic nematode *Meloidogyne javanica*. *Parasitology* **58**, 377–91.

WARD, K. A. & FAIRBAIRN, D. (1972). Chitinase in developing eggs of *Ascaris suum* (Nematoda). *Journal of Parasitology* **58**, 546–9.

WHARTON, D. A. (1979*a*). The structure of the egg-shell of *Aspiculuris tetraptera* Schulz (Nematoda: Oxyuroidea). *Parasitology* **78**, 145–54.

WHARTON, D. A. (1979*b*). The structure and formation of the egg-shell of *Syphacia obvelata* Rudolphi (Nematoda: Oxyurida). *Parasitology* **79**, 13–28.

WHARTON, D. A. (1980). Nematode egg-shells. *Parasitology* **81**, 447–63.

WHARTON, D. A. & JENKINS, T. (1978). Structure and chemistry of the egg-shell of a nematode (*Trichuris suis*). *Tissue and Cell* **10**, 427–40.

WILSON, P. A. G. (1958). The effect of weak electrolyte solutions on the hatching of eggs of *Trichostrongylus retortaeformis* (Zeder) and its interpretation in terms of a proposed hatching mechanism of strongyloid eggs. *Journal of Experimental Biology* **35**, 584–601.

Parasitology (1981), **83**, 639–650
With 1 figure in the text

Haemonchus in India

M. L. SOOD
Department of Zoology, Punjab Agricultural University, Ludhiana 141004, *India*

(*Accepted* 29 *April* 1981)

SUMMARY

Haemonchosis is an acute problem in India. Three species of *Haemonchus* occur. These are *H. contortus* (Rudolphi, 1803) which occurs in sheep, goat, cattle, buffalo and other ruminants; *H. longistipes* Railliet & Henry, 1909 occurring in camels and *H. similis* Travassos, 1941 in cattle. Seasonal fluctuation of *H. contortus* occurs, with infection being more frequent in autumn > summer > winter > spring, and infection in sheep and goats has been reported to be 100 %. Various aspects of haemonchosis have been investigated. However, the fundamental and the most challenging areas in *Haemonchus* research remain almost untouched. In order to forecast the development of haemonchosis, epidemiological studies should be undertaken on the prevalence, distribution and seasonal incidence of the worm to understand the effects of climatological factors like rainfall, humidity and temperature. The sparse and inadequate studies of metabolism need to be extended to other areas of *Haemonchus* biochemistry to provide additional possibilities for exploration of host–parasite differences. Clinical haemonchosis in sheep, goats and other ruminants should be studied in order to calculate losses caused by mortality and the cost of anthelmintic medication. Most anthelmintics are expensive, and are not easily available to our farming community, therefore a search for many more indigenous anthelmintics (such as Wopell, Krimos, Sonex etc) must be made. Detailed knowledge of the mode of action and route of entry of anthelmintics at the molecular level could also be beneficial, particularly when the strains of *H. contortus* have been identified as resistant to some anthelmintics. The prospects of producing a vaccine against *H. contortus* should be explored. Attempts in this field have already been made by overseas workers (see Clegg & Smith, 1978) and in India too, work has been undertaken at Kerala (Sathianesan, personal communication). Therefore, because of its economic importance and wide distribution, *H. contortus* provides fascinating research material particularly in the warmer regions of the world.

INTRODUCTION

Haemonchus spp. (Nematoda: Trichostrongylidae) are blood-suckers and are important parasites of domestic ruminants. *H. contortus* (Rudolphi, 1803) is the prime species occurring in the mucosa and contents of the abomasum of sheep, goats, cattle and numerous other ruminants in many parts of the world, being especially important in areas with a hot, moist climate. In India, it is considered to be the most important intestinal parasite of sheep and goats.

In a sub-tropical country like India, which has variable climatic conditions, in

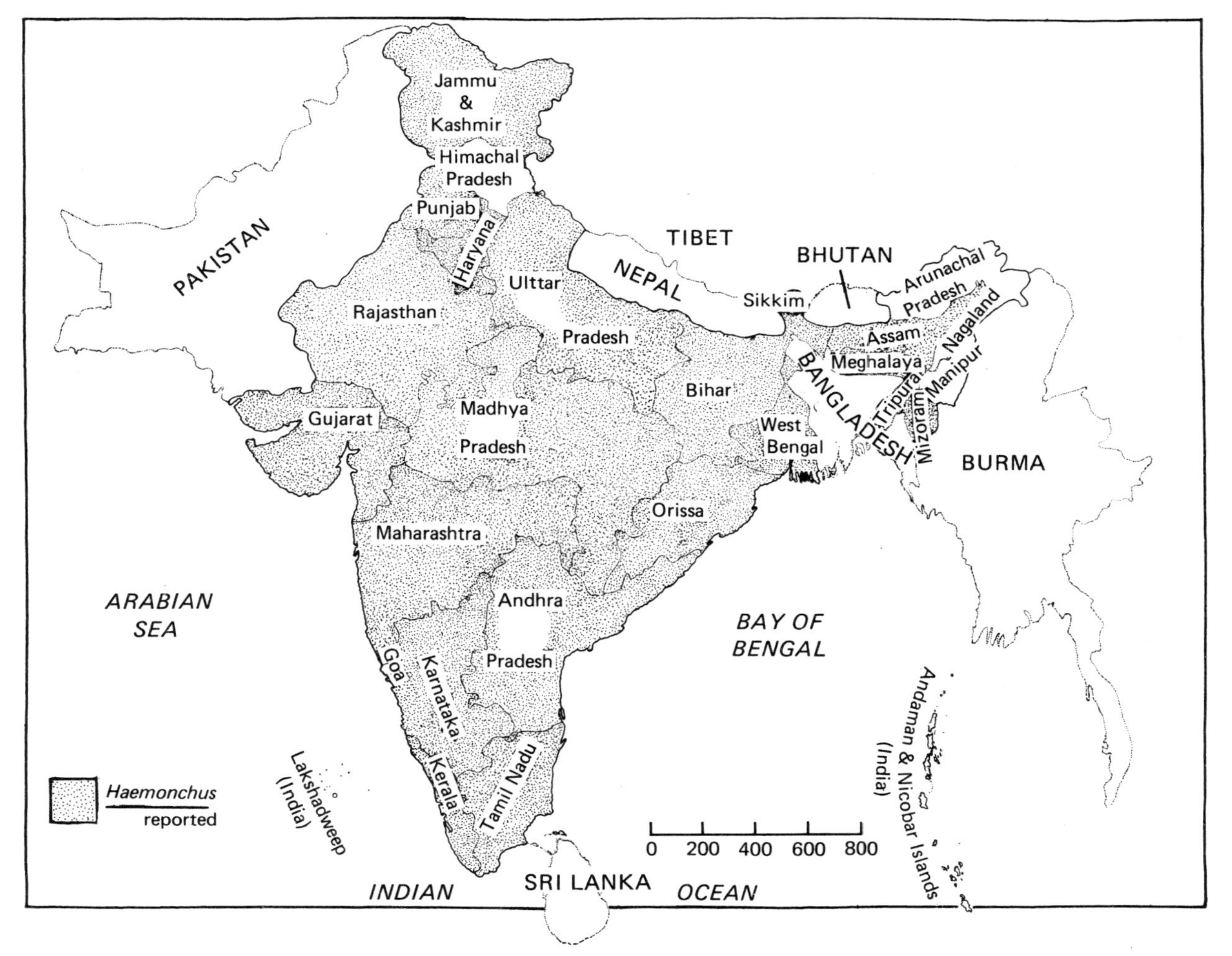

Fig. 1. A perusal of the literature reveals that *Haemonchus* is prevalent throughout India. The figure shows the regions of India where *Haemonchus* has been reported (not to any scale).

addition to malnutrition and low lying, imperfectly drained unhygienic areas, domestic mammals are invariably exposed to parasitic infections. A perusal of the literature reveals that *Haemonchus* is prevalent throughout India (Fig. 1). The importance of haemonchosis in the domesticated ruminants in India was recognized as early as 1938 by Srivastava. However, much of the work is related to isolated reports on the morphology of the worm with little information being available on its biology and pathogenicity or on chemotherapeutic aspects. The main objective of the present work is to bring together the primary information available on the various aspects of '*Haemonchus* in India' to provide a better understanding of the relationship between *Haemonchus* and ruminants. This will enable us to evolve effective prophylactic measures and to develop and evaluate chemical control methods.

MORPHOLOGY AND DISTRIBUTION

The genus *Haemonchus* was suggested by Cobb in 1898 for *Strongylus contortus* described by Rudolphi in 1803. In india, the following 6 *Haemonchus* forms have been reported.

(1) *Haemonchus contortus*

H. contortus (Rudolphi, 1803) Cobb, 1898 has been reported by numerous workers to be present in the abomasum, intestine and duodenum of cattle (*Bos indicus*), sheep (*Ovis aries*), goats (*Capra hircus*), buffaloes (*Bos bubalus*), elephants, markhor (*Capra falconeri*), chital or spotted deer (*Axis axis*), swamp deer barasingha (*Cervus duvauceli*), Indian antelope or blackbuck (*Antilope cervicapra*), Nilgiri tahr (*Hemitragus hylocrius*) and freshwater fish *Wallago* (= *Wallagonia*) *attu*. *H. cervinus* Baylis & Daubney, 1922 which was described for several females in spotted deer (*Cervus axis*) is a synonym for *H. contortus* (Baylis, 1936; Gibbons, 1979).

On the basis of pathogenicity and morphological characters such as body size and structure of the spicules, *H. contortus* occuring in sheep and goats has been assumed to occur as different strains (Rao, Sundararaj & Rahman, 1968). However, this hypothesis needs experimental evidence. I agree with Gibbons (1979) in considering *H. contortus* var. *kashmirensis* Fotedar & Bambroo, 1965 as a *species inquirenda*, since from the description and drawings it is difficult to determine its status.

In the females of *H. contortus*, 3 phenotypes have been reported; these are linguiform, knobbed and smooth. Sood & Kaur (1976) reported a greater percentage of knobbed, less in linguiform and least of smooth forms from Ludhiana in northern India. This proportion differs consistently from that recorded for *H. contortus* populations in sheep from Bangalore, southern India where Rao & Rahman (1967) and Rao *et al.* (1968) found large numbers of linguiform types, less of knobbed and a total absence of smooth types. The proportions of phenotypes reported from Ludhiana, as mentioned above, are similar to those in populations studied by Ghafoor & Rao (1970) in Maharashtra, western India.

Linguiform morphs appear to predominate in less temperate areas and smooth morphs may be better adapted to temperate areas with shorter pasture seasons. If this hypothesis is valid, then since northern India is more temperate than

Table 1. *Vulval flap formula of* Haemonchus *spp. recorded from different geographical regions of India*

	Type host	Locality	*Haemonchus* sp.	Reference	Percentage occurrence		
					Linguiform	Knobbed	Smooth
1	Sheep(?) goat	Cuttack district, Orissa	*H. contortus* var. *utkalensis*	Das & Whitlock (1960)	20–55	40–60	Under 20
2	Sheep	Bangalore	*H. contortus bangalorensis*	Rao & Rahman (1967)	67·9	32·1	0
3	Sheep and goat	Bangalore	*H. contortus*	Rao *et al.* (1968)	71·98	28·02	0
4	*Bos indicus*	Bombay, Poona and Bhir	*H. similis*	Rao & Ghafoor (1968)	96·32	3·16	0·52
5	Sheep and goat	Bombay, Poona and Bhir	*Haemonchus* sp.	Ghafoor & Rao (1970)	25·71	73·33	0·96
6	Sheep	Tirupati	*H. contortus* and *H. bispinosus*	Padmavathi *et al.* (1971)	26·44	56·74	0
7	Goat	Ludhiana	*H. contortus*	Sood & Kaur (1976)	24·6	56·94	18·47

southern India, a relative increase in the number of smooth-type females and a decrease in the linguiform females should occur on passing from southern to northern India (Table 1). The studies on *H. contortus* populations in Maharashtra, western India, also agree with this hypothesis since Maharashtra is somewhat to the north of Bangalore. The Indian record of female variants of *H. contortus* reported to date includes all the Roberts types (from A to N) except type H. In addition to the 13 variants, Sood & Kaur (1976) also recorded and designated 4 new variants as Roberts types A_1, C_1, I_1 and I_2.

Although the season has a significant effect on the distribution of the 3 phenotypes, linguiform sub-groups show no seasonal variation (Sood & Kaur, 1976). The 3 main phenotypes of *H. contortus* and their variants are all similar in body length, but are significantly different in the position of the vulva. Each phenotype has the usual 3 layers of body wall, the cuticle, hypodermis and the muscle layer in the vulvar region (Sood & Kaur, 1977).

Vulval configurations in *H. contortus* populations are taxonomically important. *H. contortus* populations in sheep (?) and goats from the Cuttack district, Orissa were designated var. *utkalensis* by Das & Whitlock (1960). These authors did not believe these populations to be a different sub-species since they did not show a constant proportion of the 3 phenotypes. Rao & Rahman (1967) gave the name *H. contortus bangalorensis* to forms collected from sheep slaughtered at Bangalore. The observations of Padmavathi, Reddy & Venkataratnam (1971) that both the knobbed and linguiform females breeding true to type, belong to 2 different species namely *H. bispinosus* and *H. contortus*, have already been questioned by Sood & Kaur (1976) since progeny were studied for only 1 generation, and the genotypes of the parents as well as the progeny were unknown.

H. contortus bangalorensis and *H. contortus* var. *utkalensis** have been considered as synonyms for *H. contortus* (Gibbons, 1979). Daskalov, Komandarev & Minhov (1972) also oppose the view of sub-speciation in *H. contortus* on the basis of vulvar

* Misspelled as *uktalensis* by Gibbons (1979).

configurations. However, immunodiffusion studies carried out with different morphological types of female *H. contortus*, have indicated that the 3 phenotypes are different from each other (Sood & Kapur, 1981). These serological studies have also been supported by electrophoretic studies (Sood & Kapur, 1980).

(2) *Haemonchus bispinosus*

H. bispinosus (Molin, 1860) Railliet & Henry, 1909 has been reported in sheep, goats and deer (*Cervus axis*).

Although *H. bispinosus* has been considered as a valid species (Sahai, 1966; Sahai & Deo, 1966; Padmavathi *et al.* 1971; Sahai & Sinha, 1979), Levine (1968) is uncertain about this species. I agree with Gibbons (1979), in considering *H. bispinosus* to be a synonym for *H. contortus*.

(3) *Haemonchus bubalis*

H. bubalis Chauhan & Pande, 1968 has been reported in the abomasum of the water buffalo (*Bubalus bubalis*). Both the adults and the infective larvae of *H. similis* and *H. bubalis* are very similar and I, therefore, agree with Gibbons (1979) in considering *H. bubalis* to be a synonym for *H. similis*.

(4) *Haemonchus longistipes*

H. longistipes Railliet & Henry, 1909 occurs both in camels (*Camelus dromedarius*) and sheep. Dutt & Sahai (1966) maintain that the *H. longistipes* reported by Bhatia (1960) was actually *H. bispinosus*.

(5) *Haemonchus placei*

H. placei (Place, 1893) Ransom, 1911 has been reported from cattle and buffalo. Sahai & Deo (1964) proposed that *H. placei* should be considered a synonym for *H. bispinosus*. According to Levine (1968), due to the poor original description, a comparison of the form from the original host deer *Mazama nana* in Brazil with that from cattle and sheep is necessary before a decision can be made. Rao & Hiregauder (1962) proposed that the goat form of *Haemonchus* should be considered as *H. placei* and the cattle form *H. contortus* but this is contrary to the proposals of other workers (Gibbons, 1979).

The *H. placei* reported by Rao & Hiregauder (1962) was thought to be *H. bispinosus* by Dutt & Sahai (1966) on the basis of the study by Sahai & Deo (1964). I agree with Trach (1975) and Gibbons (1979) in considering *H. placei* to be a synonym for *H. contortus*.

(6) *Haemonchus similis*

H. similis Travassos, 1914 has been reported in cattle and sheep. Shah & Pandit (1959) also observed that infections of *H. similis* occurred more often than *H. contortus* in sheep, which is contrary to the observations of other workers.

The morphological variants (linguiform type) reported (Rao & Ghafoor, 1968) in female *H. similis* include group I, with linguiform appendage (96·32%) (Types

A, B, C, D, E, F, G and H), group II with knobbed appendage (3·16%) (Types I and J) and group III without appendage (0·52%).

BIOCHEMISTRY

There is increasing interest in the physiological and biochemical aspects of nematodes, and recently in India, some biochemical studies have been undertaken on the adults of *H. contortus in vitro*. Reference to carbohydrate metabolism, in particular, glucose utilization and lactic acid production (Chopra & Premvati, 1977) and consumption of stored glycogen (Premvati & Chopra, 1979) indicates that such studies have been neglected in the past. Histochemical studies on the body wall (Sood & Kalra, 1977) and the gut (Sood & Sehajpal, 1978) also help in understanding the biochemical phenomenon of the parasite. Such studies on *Haemonchus* metabolism need to be extended to different stages of the life-cycle under controlled experimental conditions. The available biochemical information, restricted only to *H. contortus*, needs to be extended to other species such as *H. longistipes* and *H. similis*.

Inorganic elements detected in ashed *H. contortus* analysed by spark-emission spectroscopy are present in the following order of concentration: P > Zn > Ca > Fe > Mg > Cu > Mn > B > K in males and P > Ca > Zn > Fe > Mn > Mg > Cu > K in females. B was not detected in females (Sood & Kapur, 1980).

IMMUNOLOGY

Immunological changes in the spleen of rabbits in response to antigens of female *H. contortus* have been observed (Sood & Kapur, 1981). Hyperplasia of reticulo-endothelial cells, duplication of red pulp and haemosiderin pigment were also observed. A few eosinophils and neutrophils were present in the cortical region.

DEVELOPMENT AND CYTOLOGY

It is important to investigate the effect of anthelmintics on the formation of germ cells so that the proliferation of the parasite can be checked in its initial stages of development. Before such attempts can be made, studies on spermatogenesis and oogenesis need to be undertaken.

The structure and development of spermatogonia, spermatocytes (primary and secondary), spermatids and spermatozoa have been studied in *H. contortus* using light microscopy. The spermatozoa are elongated, narrow, smooth-walled and bluntly pointed at both ends. Mature spermatozoa are 5–7 μm long, non-flagellate and lack a typical acrosome (Sood & Walia, 1980). The structure and development of oogonia and oocytes have also been studied using light microscopy (Sood & Walia, unpublished observations). The mature oocytes are large cells measuring 16–20 μm × 5–6 μm. They are arranged around a central cylindrical strand called the rachis, which gives a vacuolated appearance due to the removal of lipids washed away during dehydration. Goswami (1976*a*, *b*; 1977) carried out karyological studies on *H. contortus* and observed that chromosomes appeared as small rods and $2n = 11$ (♂) and $2n = 12$ (♀) and the sex mechanism was XO/XX.

Table 2. *Rate of development of* Haemonchus contortus *eggs at different constant temperatures*

Temperature (°C)	Time required (h)	
	For eggs to hatch	For hatched larvae to reach infective stage
10	120–144 (132)	312–360 (316)
20	48–54 (50)	116–134 (122)
25	27–29 (28)	98–106 (102)
30	24–25 (24)	83–96 (90)
34	16–20 (18)	81–93 (84)
37	15–21 (17)	72–90 (80)
40	14–20 (16)	—

LIFE-HISTORY

The life-cycle of *H. contortus* (like other trichostrongyles) consists of a parasitic stage in which the worm lives in the abomasum of its host, and a free-living stage in which eggs pass in the faeces, hatch in the open and larvae develop to the infective stages after 2 moults. Under favourable conditions, *H. contortus* completes its life-cycle in 21 days.

The effects of temperature on survival and development of the free-living stages of *H. contortus* have been studied both by faecal and by agar-culture methods. The ova hatch and the infective larvae develop at temperatures ranging from 10 to 37 °C, but a temperature of 40 °C is lethal after 20 h. The time required for eggs to hatch and for the hatched larvae to develop into infective stages (Jehan & Gupta, 1974) is given in Table 2. Misra & Ruprah (1973*b*) considered that the optimum conditions for the development of *H. contortus* eggs into infective larvae were 25–30 °C, with a pH range from 6·5 to 8·5 and relative humidity between 70 and 85 %. Jehan & Gupta (1974) also considered 30 °C to be the optimum temperature. Significant variations occur in the body length of the infective larvae when they are developed at different temperatures (Sood & Kaur, 1975). The pre-patent period of *H. contortus* of sheep origin, in lambs, has been found to be 14–21 (17) days (Raisinghani & Singh, 1977). Misra & Ruprah (1971) found the pre-patent period of sheep *Haemonchus* to be 14–18 days in lambs and 18–22 days in buffalo calves and that of buffalo larvae to be 26–28 and 28 days respectively.

Under gamma-radiation, the survival of the eggs is inversely proportional to the exposure (Premvati, 1965). The survival period of infective larvae of *H. contortus* in the laboratory under various constant temperatures has been observed in faecal culture, in tap and aquarium waters, physiological salt solution and a filtrate of boiled host faeces in water with 1 : 10 dilution. In faecal cultures, the larvae survive for 198–205, 35, 18, 5, 1 and 0 days at 0, 30, 35, 40, 45 and 50 °C respectively (Tripathi, 1977). However, these results are quite at variance with those of Misra (1978) who found the survival period to be 252, 280, 133, 63 and 7 days at 5, 10, 20, 30 and 40 °C respectively. This variation may be attributed to different culture conditions. The results of survival in tap water (Jehan & Gupta, 1974; Sood & Kaur, 1975; Tripathi, 1977) are fairly constant but slight variations may be attributed to the effect of submergence in water (Tripathi, 1969b). The similar survival period of the infective larvae may indicate existence of only one physiological strain. The larvae survive for 140, 110, 77, 40, 33, 3 days and 3·5 h at 25, 30, 34, 37, 40, 45 and 50 °C, respectively, but at room temperature they survive for 77 days during summer (24–32 °C) and 187 days during winter (6·5–29·5 °C) (Sood & Kaur, 1975). In aquarium water, the larvae survive for 12–13 weeks at room temperature (Sathianesan & Peter, 1977) but in physiological saline solution and in boiled host faeces-water solution they survive for only 10 days (Bali & Fotedar, 1974).

PATHOGENESIS

Haemonchosis is one of the important diseases of domestic livestock causing serious economic losses. The principal effect of *Haemonchus* infection is due to its habit of blood sucking. Both 4th-stage larvae and adults are vigorous blood suckers. Migration of larvae in the abomasal wall and attachment of adults causes abomastis.

In India, ulceration and congestion of the abomosal mucosa have been reported (Bhatia, 1960; Misra & Ruprah, 1968). These authors have also described other microscopical changes in the mucosa. Purohit & Lodha (1958) observed ecchymoses of the mucosa due to *H. longistipes* infection in camel.

Haematological changes in haemonchosis have been studied by many workers. Most of the authors agree that this infection causes anaemia due to loss of erythrocytes. Hypoproteinaemia (Shastri & Ahluwalia, 1972) and decrease in serum iron levels (Mahanta & Roychoudhury, 1978) have also been reported.

The presence of *H. contortus* in the abomasum appears to interfere with the digestion and absorption of proteins, calcium and phosphorus. But, whether this is due to the damage to the mucosa and its digestive functions by the worm or to the effect of toxic substances produced by the worms has yet to be determined. Increase in the gastric acidity and plasma pepsinogen levels have been reported by certain workers. Increased susceptibility to *Haemonchus* infection due to deficiencies of vitamin A and calcium in goats has also been reported (Kumar & Deo, 1970).

CLINICAL SYMPTOMS

The clinical symptoms of haemonchosis have been summarized by Christopher (1976) and Misra (1977). These and other reports describe the general symptoms. The primary symptom is a microcytic and hypochromic type of anaemia. Other effects include loss of body weight, weakness, emaciation, loss of wool and rough hair coat, paleness of all the visible mucous membranes, change of temperature, coughing and constipation followed by diarrhoea. In more severe and chronic cases, there may be a bottle-jaw condition and swelling of the lower abdomen. At times, discharge from the eye is also seen and the infected animals may develop a habit of earth, wool or hair eating.

DIAGNOSIS

Haemonchosis in ruminants can be diagnosed by identifying 1st- and 3rd-stage larvae after cultivation of the eggs, but a skin intradermal test for *H. contortus* has also been found to be reliable. A difference in the diameter of 10 mm or more between antigen and saline wheals shows a positive reaction (Sastry & Ahluwalia, 1972; Bali & Singh, 1978).

EPIDEMIOLOGY

The reports of seasonal fluctuations of *H. contortus* based mainly on the study of adult worms agree broadly that seasonal variation is in the order autumn > summer > winter > spring, although at times, a higher incidence has been reported in winter rather than in summer. The work of Misra & Ruprah (1973*a*, *c*) on the influence of temperature on development and survival of larvae on herbage is also related to the higher level of infection in autumn and summer.

The percentage infection of *Haemonchus* spp., mainly that of *H. contortus* (separately or in conjunction with other nematode species) has been observed by many workers in many parts of India. The infection of *H. contortus* in sheep and goats has been reported to be as high as 100%.

The infection of animals with *Haemonchus*, like other trichostrongyles, takes place by ingestion of 3rd-stage infective larvae after grazing on infected pasture or ingestion with feed or water. Little information is available on the effect of environmental conditions on the distribution of *Haemonchus* larvae on pasture. Attempts have been made to correlate larval survival with climatological data. Thus, Misra & Ruprah (1974) observed that either very low temperatures (5–9 °C) with high relative humidities (91–100%) or very high temperatures (40–45 °C) with low relative humidity (5–25%) decrease the laval population. Moderate temperatures (20–29 °C) and moderate relative humidities (51–81%) cause an increase. Faecal culture studies on infective larvae of *H. contortus* at room temperature also indicate that their presence is directly proportional to relative humidity and temperature (Tripathi, 1970) but changes in relative humidity have a greater effect on the upward migration of larvae than changes in temperature (Misra & Ruprah, 1972*a*).

Significant variations in the distribution of infective larvae on the grass *Cynodon dactylon* and the soil of experimental pots have been observed in winter and more larvae were recovered from grass than from soil (Misra, 1972). The larvae survive for 182, 154, 91 and 63 days in experimental grass pots in autumn, winter, spring and summer respectively (Misra & Ruprah, 1972*c*).

Diurnal and seasonal variations of *Haemonchus* larvae on naturally infected pastures show that they are present in the order morning > evening > noon (Misra & Ruprah, 1972*b*). Direct sunlight has been shown to be more lethal to the larvae than shade or diffused light (Tripathi, 1977), and they show no phototropic activity (Tripathi, 1969*a*). Both field and laboratory experiments on their longevity show that maximum mortality occurs in sandy soils and minimum mortality in loamy soils (Tripathi, 1974).

TREATMENT

Haemonchosis is a common problem in India and anthelmintic information on clinical pathological features of the disease is still very limited. It would, of course be an important landmark in animal husbandry research if *Haemonchus*-resistant breeds (as advocated by Bailey (1978)) were available for use in developing countries. However, until such breeds are available, various prophylactic measures

Table 3. *Anthelmintics effective against haemonchosis in India*

Group	Anthelmintic	Dosage/body wt	Animal	Remarks
1	*Phenothiazine* (Phenovis-ICI)	500–600 mg/kg	Sheep and goat	Also has some effect on larvae.
		50–80 g/animal	Cattle and buffalo	Strains of *H. contortus* that developed complete resistance against phenothiazine (20 g) have been reported (Varshney & Singh, 1976).
2	*Methyridine* (Mintic, Promintic-ICI)	200 mg/kg s.c.	Sheep, goat and cattle	—
		1 ml/4·5 kg s.c.	Camel	—
3	*Naphthalophos* (Rametin, Maretin-Bayer)	50 mg/kg	Sheep and buffalo calves	Effective against immature forms too.
4	*Trichlorphon* (Neguvon-Bayer)	50 μg/kg in 10% aqueous suspension	Sheep	—
5	*Morantal tartrate* (Banminth-II-Pfizer)	—	—	Effective in *in vitro* trials against infective larvae at 100 μg/ml.
6	*Pyrantal tartrate* (Banminth-Pfizer)	25 mg/kg	Goat	—
7	*Tetramisole hydrochloride* (Nilverm-ICI)	15 mg/kg	Sheep, goat and camel	Also effective against larval stages.
		12·5–15 mg/kg	Cattle and buffalo	Also has a lethal effect on eggs at 400 p.p.m.
8	*Thiabendazole* (Thibendole-MSD)	50 mg/kg	Sheep and goat	Has a lethal effect on eggs at 500 p.p.m.
		50–100 mg/kg	Camel, cattle and buffalo	Strains of *H. contortus* that developed partial resistance against thiabendazole (3 g) have been reported (Varshney & Singh, 1976).
9	*Parbendazole* (Helatac-SKF)	30 mg/kg	Sheep	—
		20 mg/kg	Nilgiri Tahr (*Hemitragus hylocrius*) kids	—
10	*Fenbendazole* (Panacur-Hoechst)	5 mg/kg	Goat	Also effective in *in vitro* treatment against infective larvae at 900 μg/ml.
11	*Nitroxynil* (Trodax-M & B)	12·5 mg/kg s.c.	Sheep	Effective against immature forms too.
12	*Thiophanate* (Nemafax-M & B)	50 mg/kg	Calves	Has paralysing effect on *H. contortus* infective larvae at 50–100 μg/ml.

must be used judiciously. In India, the chemotherapeutic effectiveness of anthelmintics against *Haemonchus* eggs, larvae and adults have been studied both *in vivo* and *in vitro*. The effect of indigenous herbs (Roychoudhury, Chakrabarty & Dutta, 1970; Sharma, Bahga & Srivastava, 1971) and indigenous compounds (Kumar, Sinha & Sahai, 1973; Srivastava, Chhabra & Bali, 1980) has also been studied. A mixture of copper sulphate and a weak infusion of tobacco has also been successfully used (Bhalerao, 1934).

Various anthelmintics have proved to be effective against haemonchosis (Table 3). However, only a few have been recommended for use in clinical cases and some are not even easily available in our country. In addition, other anthelmintics have been tried. These include, clioxanide (Tremerad-Parke Davis), bephenium (Alcopar-Burroughs Wellcome), disophenol (Ancylol-Cyanamid), coumaphos (Asuntol-Bayer) etc. A mixture of coumaphos and phenothiazine has also been successfully tried in cattle, buffalo and camels. Morphological and histochemical changes in the body wall and intestine of *H. contortus*, caused by *in vitro* treatment with thiabendazole, morantel tartrate, tetramisole hydrochloride and piperazine hexahydrate at concentrations of 10 μg/ml and 30 μg/ml have also been studied (Sood & Kaur, 1981).

Among the indigenous herbs, a liquid extract of *Paederia foetida*, given 3 times every 2 days to calves was 100% effective against *Haemonchus*. The dose was 4–100 oz for 30–300 lb calves. *In vitro* studies on *H. contortus* in goats have shown that a variety of plants have a medicinal effect. These include *Cucurbita pepo* (seed part, 1 in 50), *Calotropis gigantica* (leaves, 1 in 50), *Juglans regia* (bark, 1 in 25), *Momordica charantia* (seed, 1 in 50), *Musa paradisiacea* (root, 1 in 25), *Mangifera indicra* (kernal, 1 in 25) and *Scindapsus officinalis* (fruit, 1 in 50).

Among the indigenous compounds, Wopell-R (Indian Herbs Research and Supply Company, Saharanpur, U.P.), when given orally at a dose of 20 g/kid proved satisfactory. Similarly, Krimos (Bhartiya Bootee Bhawan, Saharanpur, U.P.), another indigenous anthelmintic, has also shown promising results at a dose of 6 g/kid. Sonex (Animal Research Centre, Chirawa, Jaipur, Rajasthan) has proved to be an effective indigenous anthelmintic at a dose of 3 g/kid.

PROPHYLAXIS

The objective of control measures is to break the life-cycle of the parasite at one or more vulnerable points. The various prophylactic measures have been suggested by Mohan (1954), Pachalag, Chattopadhyay & Patil (1971) and Christopher (1976). These and other reports suggest drenching of animals before the onset of the rainy season and subsequent drenchings as a preventive measure to eliminate the build up parasitic population. Better management including proper and immediate disposal of dung from sheds and byres, avoiding overstocking of animals, their adequate feeding on a well-balanced diet, pasture rotation and general sanitation is quite important.

I am grateful to Drs R. I. Sommerville of the University of Adelaide, G. C. Srivastava and V. R. Parshad of Punjab Agricultural University, for help in the preparation of this manuscript.

REFERENCES

BAILEY, W. S. (1978). Veterinary medicine and comparative medicine in international health opportunities for improving human health and welfare in tropical countries. *American Journal of Tropical Medicine and Hygiene* **27**, 441–65.

BALI, H. S. & FOTEDAR, D. N. (1974). Studies on juvenile stages of some helminth parasites of Kashmir sheep. *Journal of Research* **11**, 329–43.

BALI, M. K. & SINGH, R. P. (1978). Intradermal reaction for diagnosis of *Haemonchus contortus* in sheep and goats. *Indian Journal of Parasitology* **2**, 69–70.

BAYLIS, H. A. (1936). The fauna of British India, including Ceylon and Burma. In *Nematode*, vol. 1, pp. 408. London: Taylor and Francis.

BHALERAO, G. D. (1934). The common worms of sheep and goats in India and their control. *Agriculture and Live-stock in India* **4**, 655–69.

BHATIA, B. B. (1960). On some of the bursate nematodes in abomosal infections of Indian sheep. *Indian Journal of Helminthology* **8**, 80–92.

CHOPRA, A. K. & PREMVATI, G. (1977). Glucose metabolism and lactic acid production in sheep nematodes. *Indian Journal of Parasitology* **1**, 93–6.

CHRISTOPHER, J. (1976). Controlling stomach worm in sheep. *Indian Farming* **26**, 34–6.

CLEGG, J. A. & SMITH, M. A. (1978). Prospects for the development of dead vaccines against helminths. *Advances in Parasitology* **16**, 165–218.

DAS, K. M. & WHITLOCK, J. H. (1960). Subspeciation in *Haemonchus contortus* (Rudolphi, 1803). Nematoda: Trichostrongyloidea. *Cornell Veterinarian* **50**, 182–97.

DASKALOV, P., KOMANDAREV, S. & MINHOV, L. (1972). Protein fractions in different morphological *Haemonchus contortus* female types. *Bulletin of the Central Helminthological Laboratory, Bulgarian Academy of Sciences* **15**, 57–67.

DUTT, S. C. & SAHAI, B. N. (1966). Redescription of *Haemonchus longistipes* Railliet and Henry, 1909 and *H. bispinosus* (Molin, 1860) with remarks on the taxonomic status of *H. placei* (Place, 1893) (Nematoda: Trichostrongylidae). *Indian Journal of Helminthology* **18**, 104–13.

GHAFOOR, M. A. & RAO, S. R. (1970). Studies on the females of *Haemonchus contortus* (Cobb, 1898) from Indian sheep and goats. *Indian Journal of Animal Sciences* **40**, 322–9.

GIBBONS, L. M. (1979). Revision of the genus *Haemonchus* Cobb, 1898 (Nematoda: Trichostrongylidae). *Systematic Parasitology* **1**, 3–24.

GOSWAMI, U. (1976*a*). Chromosomal studies during cleavage divisions in ten species of nematodes. *Research Bulletin of the Punjab University* **27**, 119–20.

GOSWAMI, U. (1976*b*). Chromosomes during fertilization in nematodes. *Research Bulletin of the Punjab University* **27**, 217–18.

GOSWAMI, U. (1977). Karyological studies in fifteen species of parasitic nematodes. *Research Bulletin of the Punjab University* **28**, 111–12.

JEHAN, M. & GUPTA, V. (1974). The effects of temperature on the survival and development of the free living stages of twisted wireworm *Haemonchus contortus* Rudolphi, 1803 of sheep and other ruminants. *Zeitschrift für Parasitenkunde* **43**, 197–208.

KUMAR, V. & DEO, P. G. (1970). The effects of Vitamin A, protein, calcium and phosphorus deficient diet upon the natural resistance of goats to *Haemonchus* spp. *Ceylon Veterinary Journal* **18**, 119–22.

KUMAR, G. M., SINHA, A. K. & SAHAI, B. N. (1973). Anthelmintic efficacy of 'Woppell' against hook worms in dogs and stomach worms in goats. *Indian Journal of Animal Research* **7**, 78–80.

LEVINE, N. D. (1968). Nematode Parasites of Domestic Animals and of Man. Minneapolis, Minnesota: Burgess Publishing Company.

MAHANTA, P. N. & ROYCHOUDHURY, G. K. (1978). Experimental *H. contortus* infection in goat: changes in the total serum iron level. *Indian Veterinary Journal* **55**, 187–9.

MISRA, S. C. (1972). Distribution of *Haemonchus contortus* larvae on grass and soil of the experimental pots. *Indian Verterinary Journal* **49**, 665–9.

MISRA, S. C. (1977). Problems of haemonchosis in ruminants. *Indian Farming* **27**, 25–7.

MISRA, S. C. (1978). A note on *in vitro* effects of temperature on the survival of *Haemonchus contortus* infective larvae. *Indian Journal of Animal Sciences* **48**, 322–3.

MISRA, S. C. & RUPRAH, N. S. (1968). Incidence of helminths in goats at Hissar. *Journal of Research* **5**, 276–86.

MISRA, S. C. & RUPRAH, N. S. (1971). Studies on *Haemonchus* specimens of lambs and buffalo calves, a cross infection. *Orissa Veterinary Journal* **6**, 79–83.

MISRA, S. C. & RUPRAH, N. S. (1972*a*). Vertical migration of *Haemonchus contortus* infective larvae on experimental grass-pots. *Indian Journal of Animal Sciences* **42**, 843–6.

MISRA, S. C. & RUPRAH, N. S. (1972*b*). Diurnal and seasonal variation of population of *Haemonchus* larvae on pasture. *Orissa Veterinary Journal* **7**, 155–7.

MISRA, S. C. & RUPRAH, N. S. (1972*c*). Survival of *Haemonchus contortus* infective larvae on experimental grass-pots. *Indian Veterinary Journal* **49**, 867–73.

MISRA, S. C. & RUPRAH, N. S. (1973*a*). A note on *Haemonchus* population in sheep and on pasture. *Indian Journal of Animal Sciences* **43**, 666–8.

MISRA, S. C. & RUPRAH, N. S. (1973*b*). Effects of temperature, relative humidity and pH on *Haemonchus contortus* eggs. *Indian Veterinary Journal* **50**, 136–42.

MISRA, S. C. & RUPRAH, N. S. (1973*c*). Development of *Haemonchus contortus* eggs under outdoor conditions. *Indian Veterinary Journal* **50**, 231–3.

MISRA, S. C. & RUPRAH, N. S. (1974). Influence of atmospheric temperature and relative humidity on the population of *Haemonchus* larvae on pasture. *Indian Veterinary Journal* **51**, 147–8.

MOHAN, R. N. (1954). Control of helminth parasites with special reference to livestock farms. *ICAR Review Series No.* 9. *Indian Council of Agricultural Research, New Delhi*, pp. 49.

PACHALAG, S. V., CHATTOPADHYAY, S. K. & PATIL, B. D. (1971). Heavy yield from your sheep through thiabendazole drench. *Indian Farming* **21**, 51–3.

PADMAVATHI, P., REDDY, P. R. & VENKATARATNAM, A. (1971). Studies on the morphology and development of *Haemonchus contortus* (Rudolphi, 1803) Cobbold, 1898 and *Haemonchus bispinosus* (Molin, 1860) Railliet and Henry, 1909 from sheep. *Indian Veterinary Journal* **48**, 1104–11.

PREMVATI, G. (1965). Some observations on the effect of gamma radiations on the eggs of *Haemonchus contortus* (Rudolphi, 1803). *Current Science* **34**, 380.

PREMVATI, G. & CHOPRA, A. K. (1979). *In vitro* variation of glycogen content in three sheep nematodes. *Parasitology* **78**, 355–9.

PUROHIT, M. S. & LODHA, K. R. (1958). Haemonchosis in a camel. *Indian Veterinary Journal* **35**, 219–21.

RAISINGHANI, P. M. & SINGH, B. B. (1977). Experimental infection of *Haemonchus contortus* in Magra lambs with special reference to prepatent period. *Rajasthan Veterinarian* **5**, 21–3.

RAO, N. S. K. & RAHMAN, S. A. (1967). The vulval flap formula of *Haemonchus contortus* from local sheep. *Mysore Journal of Agricultural Sciences* **1**, 168–75.

RAO, N. S. K., SUNDARARAJ, N. & RAHMAN, S. A. (1968). A comparative study of *Haemonchus contortus* of sheep and goats. *Mysore Journal of Agricultural Sciences* **2**, 128–32.

RAO, S. R. & HIREGAUDER, L. S. (1962). On the occurrence of *Haemonchus placei* Ransom, 1911 in goats in India. *Proceedings of the First All India Congress of Zoology*, 1959. Part 2. *Scientific Papers*, 454–6.

RAO, S. R. & GHAFOOR, M A. (1968). Studies on eight morphological variant females of *Haemonchus similis* Travassos, 1914 (Nematoda, Trichostrongylidae) from Indian cattle (*Bos indicus*). *Indian Journal of Veterinary Science and Animal Husbandry* **38**, 471–7.

ROYCHOUDHURY, G. K., CHAKRABARTY, A. K. & DUTTA, B. (1970). A preliminary observation on the effects of *Paederia foetida* on gastro-intestinal helminths in bovines. *Indian Veterinary Journal* **47**, 767–9.

SAHAI, B. N. & DEO, P. G. (1964). Studies on *Haemonchus contortus* (Rudolphi, 1803) Cobbold (1898) and *Haemonchus bispinosus* (Molin, 1860) Railliet and Henry (1909); with a note on the synonymy of *Haemonchus placei* (Place, 1893) Ransom (1911) with *H. bispinosus*. *Indian Journal of Helminthology* **16**, 5–11.

SAHAI, B. N. (1966). Observations on embryonation of eggs and morphology of free living juveniles of two species of *Haemonchus* Cobbold, 1898 (Nematoda: Trichostrongylidae). *Indian Journal of Animal Health* **5**, 23–32.

SAHAI, B. N. & DEO, P. G. (1966). Studies on cross infection of *Haemonchus* spp. in sheep and goats with a note on their infectivity in sucking cow-calves and buffalo-calves. *Indian Veterinary Journal* **43**, 969–72.

SAHAI, B. N. & SINHA, A. K. (1979). A note on the identity of *H. bispinosus* and *H. contortus* based on genetical development. *Indian Journal of Animal Sciences* **49**, 161–3.

SATHIANESAN, V. & PETER, C. T. (1977). A detailed study on the free living larval stages of *Haemonchus contortus* Rudolphi (1803). *Kerala Journal of Veterinary Science* **8**, 205–10.

Shah, H. L. & Pandit, C. N. (1959). A survey of helminth parasites of domesticated animals in Madhya Pradesh. *Journal of Veterinary and Animal Husbandry Research* **4**, 1–10.

Sharma, L. D., Bahga, H. S. & Srivastava, P. S. (1971). *In vitro* anthelmintic screening of indigenous medicinal plants against *Haemonchus contortus* (Rudolphi, 1803) Cobbold, 1898, of sheep and goats. *Indian Journal of Animal Research* **5**, 33–8.

Shastry, K. N. V. & Ahluwalia, S. S. (1972). Changes in serum proteins of goats experimentally infected with *Haemonchus contortus*. *Indian Veterinary Journal* **49**, 470–2.

Shastry, K. N. V. & Ahluwalia, S. S. (1974). Intradermal reaction in haemonchosis in goats and sheep. *Indian Veterinary Journal* **51**, 436–8.

Sood, M. L. (1981). Immunologic changes in the spleen of rabbits in response to antigens of female *Haemonchus contortus* (Nematoda: Trichostrongylidae). *Folia Parasitologica* (in the Press).

Sood, M. L. & Kalra, S. (1977). Histochemical studies on the body wall of nematodes: *Haemonchus contortus* (Rud., 1803) and *Xiphinema insigne* Loos, 1949. *Zeitschrift für Parasitenkunde* **51**, 265–73.

Sood, M. L. & Kapur, J. (1980). Inorganic elements in the adults of *Haemonchus contortus* (Nematoda: Trichostrongylidae). *Journal of Helminthology* **54**, 253–4.

Sood, M. L. & Kapur, J. (1980). Studies on the chemical composition of phenotypes of female *Haemonchus contortus*. *Indian Journal of Parasitology* Abstracts, p. 83 Supplement.

Sood, M. L. & Kapur, J. (1981). *Haemonchus contortus*: Immunodiffusion patterns of antigens from phenotypically different females. *Experimental Parasitology* (in the Press).

Sood, M. L. & Kaur, C. (1975). The effects of temperature on the survival and development of the infective larvae of twisted wireworm, *Haemonchus contortus* (Rudolphi, 1803). *Indian Journal of Ecology* **2**, 68–74.

Sood, M. L. & Kaur, C. (1976). Studies on vulvar configurations in *Haemonchus contortus* (Rud., 1803) from goats at Ludhiana, India. *Rivista di Parassitologia* **37**, 13–33.

Sood, M. L. & Kaur, C. (1977). Morphological and histological studies on the vulvar configurations in *Haemonchus contortus* (Rud., 1803). *Folia Parasitologica* **24**, 111–15.

Sood, M. L. & Kaur, R. (1981). The *in vitro* effects of some drugs on the morphology and histochemistry of adult *Haemonchus contortus* (Nematoda: Trichostrongylidae). *Helminthologia* (in the Press).

Sood, M. L. & Sehajpal, K. (1978). Morphological, histochemical and biochemical studies on the gut of *Haemonchus contortus* (Rud., 1803). *Zeitschrift für Parasitenkunde* **56**, 267–73

Sood, M. L. & Walia, J. (1980). Spermatogenesis in *Haemonchus contortus* (Nematoda: Trichostrongylidae). *Folia Parasitologica* **28**, 83–7.

Srivastava, G. C., Chhabra, R. C. & Bali, H. S. (1980). Efficacy of some indigenous anthelmintics and Helatac against experimenal *Haemonchus contortus* infection in kids. *Journal of Research*, (*in the Press*).

Srivastava, H. D. (1938). Helminthology in relation to veterinary science. *Indian Journal of Veterinary Science and Animal Husbandry* **8**, 113–18.

Trach, V. N. (1975). Some data on some female members of the genus *Haemonchus* (Nematoda, Strongylata)]. In *Problemy Parasitologii Materially VIII nauchnoi konferentsii, parazitologov UkSSR*. Chast 2. Kiev, USSR. Naukova Dumka. 215 (Ru.).

Tripathi, J. C. (1969*a*). Observations on phototropism of infective larvae of some common gastro-intestinal nematodes of goats. *Indian Veterinary Journal* **46**, 291–4.

Tripathi, J. C. (1969*b*). Effect of submergence in water on the infective larvae of some common nematodes of goats. *Indian Veterinary Journal* **46**, 1038–43.

Tripathi, J. C. (1970). Seasonal incidence of infective larvae of *Haemonchus* species (Nematoda: Trichostrogylidae) from faecal cultures of goats. *Indian Journal of Animal Sciences* **40**, 438–43.

Tripathi, J. C. (1974). Longevity and migration of infective larvae of some common nematodes of goats in different types of soil. *Indian Journal of Animal Sciences* **44**, 104–8.

Tripathi, J. C. (1977). Effect of different temperatures on infective larvae of *Haemonchus contortus* under laboratory conditions. *Indian Journal of Animal Sciences* **47**, 739–42.

Varshney, T. R. & Singh, Y. P. (1976). A note on development of resistance of *Haemonchus contortus* worms against phenothiazine and thiabendazole in sheep. *Indian Journal of Animal Sciences* **46**, 666–8.